Scientific Proof of Miracles

Factual Evidence for Miracles in our Everyday Lives.

Burton P. Brodt

Copyright © Burton B. Brodt
All rights reserved
ISBN-13: 978-1500943202

Contents

Introduction --------------------- 5
1. Time: Beginning and End --------------------- 9
2. Space: Beginning and End ---------------------15
3. Existence of the Universe:
    Origin of Matter and Energy ---------------------19
4. Distant Galaxies and the Big Bang --------------------- 27
5. The Properties of Light --------------------- 36
6. Strange Behavior
    of Elementary Particles --------------------- 42
7. String Theory --------------------- 46
8. Gravity --------------------- 49
9. Black Holes --------------------- 55
10. The Earth's Position --------------------- 58
11. Composition of the Atmosphere & Earth's Crust ------ 62
12. Anomalous Properties of Water --------------------- 67
13. Unique Constants: Fine Tuning --------------------- 71
14. Existence of Life --------------------- 76
15. The Minolta Paradox --------------------- 84
16. The Brain --------------------- 90
17. Other Senses --------------------- 96
18. Evolution Toward a Goal --------------------- 101
19. Biochemistry --------------------- 110
20. Quantum Theory --------------------- 120
21. Altruism and Compassion --------------------- 127
22. The Convergence --------------------- 131
About the Author--------------------- 134

Acknowledgement

I want to thank my wife Gail for strongly urging me to publish this book, and for her many hours of proofreading and editing during its construction. Without her help and encouragement, this volume would not have seen the light of day.

Also, thanks to my son Phillip for his support and encouragement after reading an early draft.

Introduction

Any discussion of miracles is liable to proceed directly into religion, but with an undercurrent of regret. Most Christians, for example, believe that the age of miracles has long passed. No one walks on water or divides the loaves and fishes or raises the truly dead. Now many people are reduced to saying it was a miracle that they survived a plane crash, that God must have been looking out for them, the same God Who just killed 125 innocent others. Deep down, many people feel a little cheated because they don't get to see what they consider real miracles any more.

But in fact there are miracles right in front of our eyes, if only we know where to look.

The title of this book may sound like an oxymoron to some. Science and miracles are normally considered incompatible. The first definition of "miracle" in Webster's Unabridged is "-- an event or effect that apparently contradicts known scientific laws and is hence thought to be due to supernatural causes -- ". That's a fairly good definition, as far as it goes. And the logical question is: if something violates known scientific laws, how can science provide evidence for it? I think the answer lies in two words: "known" and "supernatural".

No competent scientist in the world will claim that science can explain all the phenomena we observe around us. Scientists work hard to understand the truth about everything from the tiniest "elementary" particles to the so-called "edges" of the universe; yet their work never ends -- it only becomes more demanding and challenging. The more they find, the less they can explain all their observations. If everything were known, there would be no need for science. Even most scientific "laws" are not known to be a true and permanent picture of reality; they remain laws only as long as they appear to work. Often they must be discarded and new laws devised. Violations of "known scientific laws" occur all around us, all the time. This book will examine some of these events.

I suggest a somewhat broader definition of "miracle" than the one in Webster's. I propose that there are four kinds of miracles, of differing strengths.

The first, and strongest, comprise facts that are, quite simply, impossible. They cannot be explained, rationalized, or understood by *any* conceivable thought processes of which the human mind is capable. It's all we can do even to contemplate them.

The second kind of miracle is an event or effect that, like Webster's definition, cannot be logically explained by "known" scientific "laws".

The third type is the result of a series of coincidences so incredible as to defy belief that they are merely fortuitous. I admit that on its face the third kind is sometimes, but not always, weaker than the first two, but in the course of this book I hope to show that there are so-called coincidences so unlikely that they easily deserve the name miracle.

Finally, the fourth kind is phenomena that may have naturalistic explanations; unfortunately, no one is sure what those explanations are.

Some scientists have a great deal of trouble with these concepts. They try so hard and sincerely to understand nature that when no known law seems to apply they develop theories to take the place of laws. Then they set out to prove or disprove the theories. So far, so good, (although it often turns out that the more difficult it is to explain something, like the origins of matter or life, the more bizarre the theories). But then if, after laboring assiduously, they find a minute bit of evidence supporting their theory, it often becomes to them a law to be proclaimed across the land, in spite of the fact that the conclusion may be a logical absurdity. In some cases, the theory becomes a "truth" without a shred of evidence, simply by being promulgated loudly and at length by enough scientists and science popularizers. I will discuss several such theories in this volume.

Another scientific failing that is occurring at the time of this writing is the use of "science" in the cause of politics. For example, the theory of man-made global warming (now called "climate change" due to some inconvenient data) is being used to advance political, environmental, and economic causes. Unfortunately, proponents are using the very unscientific terms "settled science" and "the debate is over". In real science, these terms are never appropriate or accurate.

These are rather harsh remarks, and certainly most scientists do not fit that picture. But unfortunately, those who do are often the ones most adored by the popular press, and who receive the biggest grants.

The encouraging news is that increasing numbers of prominent and respected scientists are recognizing the more we learn about how the universe actually works, the less we truly understand it. Many are coming to the conclusion that our ignorance of fundamental causes is so deep we may never be able to explain them adequately with our mundane "naturalistic" view.

Throughout the book, I hope to avoid any recourse to religious dogma or miracles, and to adhere strictly to facts, logic, and probabilities. However, I readily admit that this is not a completely objective scientific treatise. It contains many personal viewpoints, generously sprinkled with currently-held, scientifically based opinions. Also, scientists reading this book may be distressed to find their specialty grievously simplified. As a member of the scientific community myself, I expect to catch a little flak over this book, both because of the intentional over-simplification and because I seem to have deviated into heresy. To those scientists who find it offensive for a fellow scientist to question some of the accepted wisdom, I would draw their attention to the fact that the concept of heresy should never be part of the scientific rationale. And I would also ask them to stop and think: could it be that our science is missing the main point of it all? Are we being blinded by our own dogmas, foregone conclusions, and "generally accepted facts"? Are the realities of the universe trying to tell us something that would change our entire method of thinking?

    This small book will attempt to show that scientific dogmas actually explain very little of real consequence; that, in fact, they point out repeatedly that miracles are occurring all around us, if only we will see them.

1. Time: Beginning and End

In this initial chapter, we'll ask some age-old questions. When did time begin? When will it end? How could it end? Is it a fourth dimension? Can it be distorted and manipulated? Can it be traveled in both directions? Depending on the answers, time itself may be our first and most obvious inexplicable phenomenon.

Scientists like to think of time as a dimension, just like length and width and thickness. The concept of time as a fourth dimension is so universal that it has been popularized right down to the bottom of society. I have heard homeless people mention the fourth dimension. Since it is impossible to see this fourth dimension the way you can see the other three, how did it come to be considered an equivalent measure?

There are three main ways of looking at time. One is the Newtonian concept of linear, invariant time, in which a second is always a second, no matter when or where. The second is the Einsteinian theory of relativity, in which time is shortened as velocity increases. The third is that time is part of an actual *thing* called "space-time". This will be discussed further in Chapter 3.

In the Newtonian scheme, time is a dimension only because time can often be a convenient way to express distance, especially if you maintain a constant velocity. For example, if you are unfortunate enough to be starting out to drive from Philadelphia to New York on the New Jersey turnpike, you might say that New York is two hours away. That assumes you drive at some given speed. Astronomers, of course, routinely use this shorthand method to describe astronomical distances in some manageable way; it's a lot easier to say one million light years than to say 5,876,103,000,000,000,000 miles. They can do this because in a vacuum light travels at a constant speed of 186,330 miles per second.

Does all this mean that time is actually a fourth dimension? To answer that, we have to define "dimension". A dimension, as used in this context, is merely a direction. It does not pinpoint a location, even if there is only one dimension. An infinitely thin straight line is a single dimension, but a spot can be located anywhere along it. An infinitely thin flat surface is two dimensions, and not only can an object be located anywhere on the surface, but its options for that possible location are increased tremendously. In three dimensions, an object of any size and shape can be located anywhere (except in quantum physics). Not only are these three dimensions (or directions) sufficient to locate any object, but the dimensions themselves can be distorted to an infinite extent. Lines representing the three dimensional directions can be lengthened or shortened, bent, twisted, manipulated at will. Applying this picture to time, it is obvious that in Newtonian physics time does not meet the criterion for a dimension. For all practical purposes, in the everyday Newtonian world, time is invariant. It cannot be distorted or manipulated. It goes on inexorably, in one direction. Everything exists at a single moment in the line of time. What was is gone forever. You can capture an image on film or in your mind, but it is an image of a moment that will never come again.

Einstein destroyed that concept of time. His theory of Relativity states, among other things, that as an object's velocity approaches the speed of light time slows down, and that when you reach the speed of light time becomes zero. Thus the mathematical equations include time as an actual fourth dimension. But even in Relativity Theory, time does not go backward.

In spite of the hopes of science fiction (and even a few actual scientists), backward time travel is logically impossible. You can easily prove this to yourself. If you could travel back in time, you could (and very likely would) change something that happened then. If you interacted in the slightest way with any of the millions of your ancestors, even indirectly, and if that interaction changed their actions or the actions of any subsequent ancestors by the tiniest amount, the most likely outcome would be to cause you to disappear, along with most of your relatives over the ages. That's how close you came to not being born at all. And if you weren't born, of course you couldn't have traveled back in time in the first place.

Relativity gets everyone all excited about the possibility of high speed space travel in which the aging process slows, so that when the intrepid travelers return, still young, the world will have aged hundreds of years. It would be a form of time travel, except forward in time. This is all very nice, but it's nearly as absurd as backward time travel. The thing that the science fictionalists forget is that if you accept Relativity Theory, as an object's velocity increases its mass also increases. It could not reach the speed of light because the mass would be infinite. That means the impulse or force rate necessary to reach the speed of light is also infinite, and thus unreachable. But even at lass than light speed, the velocity necessary to slow time enough to see a significant difference when you returned from your trip would be a large fraction of light speed. Again, the impulse needed would be far beyond the bounds of possibility.

There have been some experiments in which hydrogen nuclei have been speeded up in a particle accelerator to some very high fraction of the speed of light, and a slight slowing of its decay rate has been measured.

This tiny effect probably qualifies as proof that time has slowed. Even so, significant forward time travel works only in books or movies.

For more about the properties of light, see chapter 5. In the meantime, what is all this leading to? The point is that time is unidirectional and, for a mass under normal conditions in the universe, nearly invariant. Some minor variations (called "relativistic") occur in satellites speeding around the earth, but their magnitude is almost negligible. Whenever you express distance as a time, whether on the New Jersey Turnpike or in intergalactic space, the only reason you can do it is that time is always essentially the same. Velocities vary, directions change, distances can be anything. But the one unchanging variable in the observable universe is the magnitude of time. It goes on and on, invisible, uncontrollable, unmeasurable except by various gadgets that we hope work at a reasonably constant rate. And if time is invariant and unidirectional, where is its beginning and where is its end?

From the viewpoints of ordinary logic and of known physical laws, the idea that time started Here or Then is poppycock. There is absolutely no hard evidence of such a start anywhere. Furthermore, if one thinks about it logically, rather than as a faith to be repeated, it is nonsense to consider that time had a start, and that before that there was no time. How could that be? What was there?

And yet, and yet -- here's where the miracle comes in. How **could** time have existed forever? Forever? That's a much-misused word. To the lover, it means as long as we live, a very short time indeed. For a man waiting for his pizza delivery: "This is taking forever". But true forever is not a period of time. It is, in fact, one of the most terrifying concepts in existence. It is even more terrible than a hall with opposing mirrors in which one's image recedes into infinity to be swallowed up, because even in the hall there is finally an end where the light darkens and the image is too small to see. But forever has no end and no beginning. It is impossible to comprehend, impossible (one must think) even to exist. To the human mind, everything must have a beginning and an end.

Yet how could it not have existed forever? What could there have been before time began? How could it possibly have had a beginning? That too is impossible to conceive, not only for the layman, but for anyone. Many scientists believe that time started with the "Big Bang". See chapter 3 for more about this theory. Suffice to say that the Big Bang explains nothing about how time could not have existed previously.

So we're left with the paradox: how could it go on forever, and how could it **not** go on forever? These concepts cannot be explained or even rationalized scientifically. In their frustration, scientists dream up universes in which time "curves" somehow until it comes back on itself. Thus everything that happens has happened before an infinite number of times, and will happen again infinitely often. Of course, the most obvious fallacy in that piece of wishful thinking is that it too presupposes infinity, except in this case a circular infinity. There is no evidence of such a thing, it makes no logical sense, and it gets us nowhere in our understanding.

Over the ages, this quandary of infinite time, of impossible beginnings and ends, has driven many thinkers half-mad, and has spawned some pretty amusing theories. One popular theory in the middle ages and even later was that the entire universe is merely a dream (often the king's dream), and that when the dreamer awakes the whole works will disappear in an instant. Similar frustration over the inability to comprehend infinity has been instrumental in developing the idea in some minds that there is no absolute truth. But there is truth, and you and I both know it. What we don't know is what that truth is.

In some ways, it is questionable whether time itself exists. There are a few respected scientists who believe it does not, that in fact the universe is actually an infinite number of "Now's", all of them static and unchanging, something like digital electronics. This sounds delusional to me. However, they make one good point. Einstein failed to develop a Unified Field Theory, that is, to reconcile astronomical equations with sub-atomic equations. But if time is omitted from the equations, they apparently fit together. In any case, whether or not time exists, whether or not it can be considered a mathematical dimension, it is not a thing or a measure at all. It is merely a concept, an abstract way of thinking about the interval between the ticks of a clock, the beat of a heart, or the revolutions of a planet around the sun. It is a useful concept, helpful in keeping track of past days and ages. We can "think back" far into the past. But when we try to think back to infinity, we cannot do it.

If the paradox of time's beginning and end do not fit the definition of miracle, nothing does. Somewhere there may be an answer, but not in our naturalistic world.

For more on this subject, see chapter 22.

2. Space: Beginning and End

In some ways, the Fourth Dimensionists are right; space and time are similar. But not as dimensions. Rather, they are similar in their invisibility, their impossibility of measurement except indirectly with gadgets that we hope correlate with what we're measuring, and particularly their paradoxes of infinity. Space supposedly is nothingness, a nothingness through which trillions of huge bundles of mass and energy are moving, an infinite nothingness that stretches out in all direction to -- nothing. Or does it? Let's look at some of the attempts recent scientists and philosophers have made to explain space.

Because of the peculiar properties of light (see chapter 5), scientists for many years believed that space must contain an invisible medium that they called "ether". Nothing else would explain how light could travel in waves. So in 1886, Albert Michaelson and Edward Morley devised an elaborate experiment to measure the ether. They build an apparatus that would detect differences in the speed of light so accurately that the effect of the earth's rotation would be seen in the velocity of light coming and going in the direction of the rotation and to and from a line perpendicular to the rotation. They ran the experiment and found much less difference than expected. Later experiments up to modern times with more accurate equipment confirmed that the difference was negligible. The velocities were about the same. It appeared there was no ether, although scientists refused to believe it for many decades. Of course, at that time Einstein had not yet shown theoretically that light velocity should be the same to all observers, regardless of the observer's velocity.

Einstein and others have stated that space is curved. They base this primarily on the fact that light bends slightly as it passes close to a massive object like a star. Of course, as will be discussed in chapter 5, light reacts to gravity just as if it had mass. If the gravity is great enough, the light is drawn into whatever is the source of that gravity, and if it cannot escape we have a black hole. The fact that light bends as it passes through a large gravitational field proves nothing about space. How can space bend if it is a nothingness? Another indication that space is much more than a nothingness will be shown in subsequent chapters.

Chapter 8 will address the miracle of gravity. Suffice to say at this point that in order to "explain" gravity to the lay public, and perhaps to themselves, scientists like to say that gravity is a bending of space in the vicinity of large bodies of mass. Thus they use gravity to explain the bending of space, and the bending of space to explain gravity. Would any professors of Logic care to comment?

There was a great controversy about whether gravity acts at the speed of light, or instantaneously, or somewhere in between. The answer, if anyone could figure it out, would say volumes about the nature of space. This important subject will be discussed in Chapter 8.

There are significant data to support the theory that the universe is expanding outward in all directions. Expanding into what? Well, space. Empty, infinite space. It has never bumped into an end point, a wall of some kind that defines the limit of space, and it never will. How could it? What would be beyond the wall? Some of the ancients used to think of the sky as the inside of a black ball containing thousands of tiny holes through which the light of Heaven shone as stars. No doubt the question arose in even their minds as to what was beyond the walls of the ball; the beauty of the theory was that they could imagine anything they liked, and no one could prove them wrong.

The irony is that many conclusions of today's scientists fall into the same category.

Just to further confuse the issue, the commonly accepted theory is that space itself is expanding (see Chapter 4). Again we ask: expanding into what? Into nothing, apparently. So if space is expanding into nothing, then how can space be nothing?

If the infinity of time is terrifying, the infinity of space is almost enough to unhinge the thoughtful mind. Even the concept of infinity itself is beyond human comprehension. The idea that space is somehow curved offered a refuge from the unimaginable thought. Perhaps it could curve back on itself and not be infinite at all. As an undergraduate, I myself wrote an extremely sophomoric (or worse) essay on the subject, in which I theorized that since we define an astronomical straight line by its relation to other cosmic bodies, then if space could curve, a straight line as we define it was not really straight, that therefore the galaxies, etc. were moving in non-straight lines relative to each other, that since we made all our astronomical measurements as if straight lines were straight, in reality the galaxies might be bending around in great cosmic circles, and if we looked out far enough we would see ourselves receding. It is a testimony to the difficulty that infinity inflicts on the scientific mind that something similar to even this asinine theory has been bandied about in some circles.

The question remains: what is space? Is it nothing? If so, how can it curve? Except for the faulty example of light bending, there are absolutely no data showing such a thing. Invoking a stray molecule or two floating around in the space between astronomical bodies gets us nowhere. Even the idea that there are enormous amounts of "missing" matter, or even anti-matter, somehow invisibly filling up the interstices of space gets us nowhere. Whatever is out there, in the unlikely event it is anything at all, the space it is in remains just that: nothing.

Or does it? The inescapable fact is that light and other forms of energy do travel as waves through space (see chapter 5 for more on this). It is also true that apparently there are huge gravitational forces throughout the universe that cannot be explained by any known bodies of matter (chapters 3 and 4). All of which begs the question: is space really nothing? Or is it packed with something invisible to us, something undetectable? Is space itself something different from the vacuum we think it is? If the answer to either of the above questions is yes, what natural law does that follow?

Does space go on forever? How can it? How can it not? Infinite space, like infinite time, cannot be explained by any scientific law or theory. It cannot even be visualized by the human mind. To that extent it is, by our definition, a miracle. And if it turns out that space is entirely different from nothing, that too does not fit our understanding and our natural laws, and it remains a miracle.

The subject of infinity will be addressed further in Chapter 4.

3. Existence of the Universe. Origin of Matter and Energy

Now we come to what I consider one of the meatiest and most obvious miracles of all: the existence of mass and energy. No longer are we discussing esoteric, invisible, mere apparent concepts like time and space. Now we're looking at something that is unquestionably real and concrete. Unless you happen to be one of those left-over folks from the Middle Ages who believes that everything is a dream, matter exists. Energy exists. In fact, there is no difference between them. Mass **is** energy. If I may digress on that subject for a moment before addressing the main topic, every sentient high school student knows that $e = mc^2$, where e is the energy content of mass m, and c is the velocity of light. (In the interests of full disclosure, I should say that the actual equation contains a velocity factor and is exact only when the mass is stationary; if it's moving at nearly the speed of light, there is a slight variation.) Of all the amazing and often correct things Einstein did, nothing can match this elegantly simple and incredibly unlikely equation. Unlike most scientific equations which need a constant (or "fudge factor" as engineers call it) to make it correct, Einstein's equation doesn't even have a constant. The energy content equals the mass times the square of the velocity of light. Period.

Wait a minute. The velocity of light? Think about that for a moment. Matter, energy, the velocity of light. What a very strange equation. What unlikely bedfellows. As discussed in chapter 5, light has a constant velocity in space, so c^2 takes the place of the constant fudge factor. And that velocity is intimately bound up in the equivalence of matter and energy. The mind reels. Chapter 5 will reel deeper into this.

But getting back to the subject, undeniably there is matter/energy in the universe. There is a lot of matter. Consider your automobile. It contains so many trillions of atoms that the earth's gravity is pulling on it with a force greater than you, your spouse, and all your children put together can overcome (unless you have many very large children indeed). Think about your house. You can't lift that no matter how many children you have. Now think about your whole town or city. Consider the entire earth, the weight of it, the mass/energy, the sheer enormity of it. Peanuts. Take the sun. It contains so much mass, mostly hydrogen, that when it formed, its own gravity squeezed it to a core pressure of three hundred billion pounds per square inch and a temperature of a million degrees Centigrade, enough to ignite a continuous thermonuclear fusion, which then raised the core temperature further to fifteen million degrees Centigrade and the pressure to 1.7 trillion pounds per square inch. The sun is only average. There are billions of stars that make the sun look like an undernourished ant. Put hundreds of billions of stars and other immense collections of matter in a group held together by its own gravity and you have a galaxy. Look up on a dark night and you see the faint glowing streak of distant stars we call the Milky Way, and that is just a portion of our own galaxy. There are hundreds of **billions** of galaxies. Obviously, the total amount of matter and energy in the known universe is so great as to defy comprehension.

    Now we come to one of the two greatest questions facing the thinking mind, a question that no honest scientist can ignore while claiming there is a naturalistic explanation for everything. Where did it all come from? How did that unimaginable amount of matter/energy originate? For that matter, how did any single atom originate? Any electron? Anything? Without an answer to that question, we know nothing, at least nothing of substance.

As an example of the difficulty this question poses for science, the famous scientist Stephen Hawking asked the same question, which I'll paraphrase. Where, he asked, did all the matter come from? His answer: from energy. And where did the energy come from? Well, he said, the universe if full of gravity, which is a negative energy. The negative energy exactly cancels out the positive energy. Thus the total amount of energy in the universe is zero. So what's the problem?

If that isn't the King's dream, what is? This man is very smart. But he obviously has no more idea where it all came from than you or I. Maybe less.

I traveled in Thailand a few years ago, and talked to a few Buddhists. Whenever I encountered a particularly intellectual person there, I would slip in the question of what he thought about the subject of originations. All of them had essentially the same answer: we don't worry about that. The universe has always been there.

In a way, their concept of infinite existence is no more difficult to accept than that of infinite time or infinite space. But in another way, their rationalization seems a weak evasion. Unlike space or time, which may be just abstract concepts, matter and energy are real. They can be seen and felt and measured. Our minds are attuned to the need for a beginning and an end. Most scientists talk about the beginning of the earth five billion years ago. Some faith-based scientists say it has existed for only ten thousand years. Either way, it pleases us to believe there was a beginning and there will be an end. Yet we seem to worry not at all about the beginning of an electron. How could there have been a beginning? How could there not?

This question has vexed mankind for as long as humans have had a human brain. Nearly all the world's religions, from the most primitive to the most sophisticated, have explained it to themselves by saying that God created everything. They admit they don't understand God or creation, but at least they have an answer that satisfies them. But most scientists are unable to accept that answer. Their creed is that everything has a natural explanation. So they search for that explanation, often with amusing results.

In about 1985, when I was a research manager at the DuPont Company's Experimental Station in Wilmington, Delaware, I attended a seminar one night in which a prominent visiting astrophysicist spoke at length on how the universe might have been formed. The theory was that an instability occurred in the nothingness of space, an instability that led to an oscillation of ever-increasing magnitude, until suddenly the entire nothingness split into equal parts of matter and anti-matter. He said since it is well-known that when matter and anti-matter come together they annihilate each other to nothing, why couldn't nothing go the other way and become something? At that point, he began to lose his audience, because matter and anti-matter do not become nothing -- they produce energy. Where did that energy come from to reverse the process? Also, people were uneasy about the original source of this wondrous instability; the speaker was pretty vague about it. But even so it was an exciting concept. We wanted to believe it. I could almost feel it: the oscillation, the sudden split, the gravitational attraction, the lighting of the stars. The strangest thing about the theory was that it almost got us back to the beliefs of the Ancients that everything was imaginary, a dream. The universe was not actually real, but was only the visual half of a split nothingness, a half that would someday revert back to -- nothing but energy. I didn't much like the idea, but it was an answer. For the first time, there was something naturalistic, even if fantastically esoteric, that might explain this basic mystery.

But then came the let-down. Someone asked: what about that anti-matter? Where is it? The astrophysicist cleared his throat and began a long, involved explanation of the extensive experiments that were carried out to find and measure the anti-matter. They were good experiments. If the anti-matter were there, the experiment would find it. Finally, someone else asked: well? Well, mumbled the speaker, it wasn't there. A silence descended over the room. I remember looking at my watch and thinking: Good grief, I've wasted two hours of my life. The meeting broke up minutes later, applause was sparse, and only a very few people went up to talk to the speaker. I almost felt sorry for him.

As I later learned, the lack of an equal amount of matter and anti-matter in the universe is one of the great unsolved mysteries in science. It is a violation of the laws of physics, as defined by quantum mechanics. See chapter 20 for more on this.

Recently, astrophysicists and others are saying that there is also missing ordinary matter out there, based on the amount of gravitational force being exerted, and that not just 50% of the universe is missing but about 95%. Less than 5% of the matter/energy in the universe is visible; 22% is invisible (or Dark) matter. The remaining 74% of the universe is believed to be Dark Energy, some kind of invisible something that permeates space throughout the universe and is pushing the galaxies apart. The physicists are beginning to think that some of it is in the form of neutrinos, strange little particles that zip through the earth and the sun as if those massive bodies were not there. Some scientists think that neutrinos have no mass, but others think they do. They appear to be thinking up facts, which they ought to be measuring instead. But in either case, that leaves most of the missing matter still undiscovered. Some say it's anti-matter, although just as in the case of the unfortunate speaker in Wilmington, they can find almost no trace of it. One wonders about the nature of a universe 95% of which can't be located by people smart enough to find a Charmed Quark or the invisible planet of a distant star. Maybe they'll find their missing matter some day, but even if they do, where are we? It still won't explain where it all came from. Furthermore, there are implications for the transmission of light (see Chapter 5).

Ah, but there is an answer to the question of origins, a theory that everyone knows about, one that daily grows stronger in the minds of both scientists and laymen as the data pour in and the press releases pour out. It is grand, it is wonderful. Everything fits. Everything is explained. This fine theory is called the Big Bang. In case you've been in Patagonia for these past 20 years, the most common scientific exposition of the Big Bang states that everything, all the matter and energy of the universe, as well as all the time and space, came out of a single microscopic speck about 13.8 billion years ago. One day, no doubt feeling rather confined, that speck expanded to the size of a cherry seed, then to a grapefruit. After that it really went to town and blasted out the entire universe in nanoseconds. What a speck! As the old-timer might say: They shore don't make specks like that no more. Some scientists say there was not even a speck; everything came out of a "singularity". A singularity is an infinitely small point, which means it is essentially nothing.

There is another version, and it will be discussed in Chapter 4.

Thousands of absolutely brilliant scientists of all disciplines have been working to verify that the Big Bang actually occurred. So far, the data seem to confirm it. The universe appears to be expanding. The galaxies probably are flying apart. They have discovered some background radiation of just the right frequency to allow them to say it came from the original explosion that sent the galaxies on their mad voyages in all directions. And recently they may have discovered some gravity waves that they say must have come from the big explosion. These people are incredibly smart and competent. Their ability to compute the presence and orbit of a distant planet from the minute wobble of the star it circles is awesome. Perhaps they are right and there was a huge explosion. If they say there was, and their data confirm it, fine. But rarely have I read that one of these geniuses has ever slapped himself on the head and said: Wait a minute. What the blazes are we talking about here? Does this theory follow any known or conceivable natural laws? Does it make sense? Is it even possible? Surely it must be obvious to anyone, no matter how deeply committed he is to naturalistic explanations of everything, that having the universe erupt out of nothing is a supernatural event. It is Creation with a capital C. Except for one three-letter word, there is absolutely no philosophical difference between it and the creationism of religious believers. Furthermore, as I will show in chapter 22, this creation event has even more implications in the struggle between science and religion.

I once challenged some of my so-called non-believing friends with this conundrum. How can you sneer at religion, I asked, and still believe in the Big Bang? How could such an explosion possibly have occurred within the framework you have erected of laws and logic? The answer was practically the same as that given by the Buddhists of Thailand. We aren't worrying about that. This is what the data say, and that's it.

Well, I beg to differ. That's not it. There's one more step. That step is to acknowledge that science has collected data which define and delineate - - a true miracle.

4. Distant Galaxies and the Big Bang

Let's look more closely at the Big Bang, assuming it actually occurred. The main theory is that prior to about 13.8 billion years ago there was nothing. No matter, no energy, no space, no time. Suddenly, there was -- everything. From a single point in the nothingness, a point the scientists call a "singularity", the entire universe rushed out in a massive explosion. The new universe expanded out from this point, carrying its own space and time. For the non-religious, thinking deeply about the foregoing does not make it any easier to conceive, and obviously most scientists have chosen not to dwell too long on the implications. But let us dwell anyway.

The other version of this event is that the universe did not come out of a singularity or infinitely small point, but that all the already-existing matter and energy of the universe was gathered for some reason in one general location that was extremely dense and hot. The universe then expanded from that mass. Let's call that version 2.

If the Big Bang occurred as advertised in either version, there should be a point in the universe from which all that matter, space, time, etc. originated. All the galaxies should be moving away from that spot, as well as from each other. Science popularizers like to compare the galaxies to spots on a balloon that is being blown up. But as a description of the universe, the balloon is a poor analogy. If it were true, all of space would be in the form of a hollow sphere, and all the galaxies and other matter would be receding from that single point in the center of the sphere. But instead, space is isotropic and homogeneous; that is, full in all directions. No matter where we look, we see billions of galaxies and trillions of stars, at all distances from 4 light years to at least 13 billion light years from our own star (also known as the Sun). What kind of explosion can produce that result? And what caused the explosion?

Let's look at that last distance -- 13 billion light years. Now that the wonderful space telescopes are available, cosmologists insist that the most distant stars they can see are about 13 billion light years away. That means the light from these stars left them 13 billion years ago. They say these stars were formed early in the life of the universe, less than 500 million years after the Big Bang. Some even used to claim that soon we may be able to look out further and see the Big Bang itself. The illogic of that is too obvious to discuss. Scientists now say that for the first 800 million years after the BB, space was opaque or foggy because all the photons, or particles of light (see chapter 5) were being knocked about by free protons and electrons, so the universe from those years is invisible. Only when the protons and electrons calmed down enough to get together and make atoms was the light free to move about and light up the universe

What's wrong with this picture? If the light took 13 billion years to reach us, that means the galaxies from which it came were 13 billion light years away 13 billion years ago. If the Big Bang occurred 13.8 billion years ago, that implies the galaxies moved 13 billion light years away from us in 800 *million* years. How did they get so far away in such a short time? They would have to travel at an average of 16 times the speed of light, relative to us. Wouldn't Einstein disapprove? Furthermore, if they continued to speed away at 16 times light speed during the 13 billion years before their light reached us, they would now be 208 billion light years away.

Astronomers will attempt to explain away this paradox by invoking the expansion of space and time. Some say that for the galaxies to move away from each other at greater than light speed does not violate any natural laws because time is being carried along with the expanding space. It seems if you accept the idea that time and space together form a concrete "thing" called space-time, you can use it to rationalize any paradox that comes along.

Another question arises: If the universe is isotropic and homogeneous, then we can see out to 13 billion light years in any direction. If the "end" of the universe were only another 800 million light years away in any direction, that would mean that we are in the exact center of the universe, occupying the very spot where the Big Bang occurred. This is another absurdity. In fact, the universe being isotropic means that no matter where in the universe you are, you see the same collection of stars in all directions.

What does this say about the size of the universe? If, no matter where in the visible universe you are located, you can never see the end of it, but can see only more stars and galaxies out to 13 billion light-years from wherever you happen to be, this would imply the universe is infinite in radius, or at least a great deal larger than the standard 13.8 light years. And in fact that is exactly what cosmologists are now saying. They estimate that after the light from those most distant visible galaxies started their long journey, the continued expansion of the universe means those galaxies are now much farther away. There are many distance estimates, including mine, the most popular of which is 46 billion light years, meaning the universe is 92 billion light years in diameter. That's still too small to explain how the oldest stars got 13 billion light years away in 13.8 billion years, but it's a start.

All of this becomes almost academic when you consider the results of recent esoteric measurements by

NASA and others to see if the universe is curved, as they have always believed, or is "flat". And they have been somewhat disconcerted to find it is, in fact, flat. That doesn't mean we are in danger of falling off the edge of the universe. It means a straight line through the universe *and beyond* is always a straight line, forever. Some scientists say that means the size of the universe is infinite, even if it doesn't all contain any matter or energy. Others are equivocal; they say maybe it's infinite, maybe not. I would guess their hesitation means they are reluctant to admit the likelihood of an infinite universe, because infinity confirms the miracle of chapter 2 and is impossible to contemplate. Furthermore, it casts doubt on the premise of the Big Bang -- that space was created in that event. Perhaps the kind of crowded space we find ourselves in, full of mysterious matter and energy, was formed as advertised. But the kind of space that really is a nothingness exists, if it can be said that nothingness "exists". And, as discussed in chapters 1 and 2, infinity is a miracle in itself.

If cosmologists are asked where in the universes is the spot where the big bang originated, they will say: "Right here, in your living room. And everywhere else in the universe." The reason they give is that at the moment of the explosion's start, all areas of the universe were together in one spot. For about one millisecond, this sounds reasonable (if you believe in that magic singularity), but further reflection casts a lot of doubt. Regardless of whether or not everything was once in the same spot, it is no longer together. We now have a spherical universe (supposedly), which means there is a central point. Everything is believed to be moving away from everything else, an observation that justifies belief in the Big Bang. So where is that starting point? Would it not be in the center of the sphere? And why is there no completely empty space in the sphere? Again, scientists fall back on the creation of space-time to explain away the paradox.

Here's another question. If the stars have all been flying apart for 13.8 billion years, why are there so many objects still so close to us? Why is the Andromeda galaxy approaching on a collision course with our Milky Way and threatening to tear apart both in about two billion years?

There are three main factors that convince most (but not all) astronomers that the Big Bang occurred. One is the expansion already mentioned. Another is a faint background radiation throughout space having an average temperature of 2.7 degrees Kelvin. Temperatures in Kelvin degrees are the same as degrees Centigrade above absolute zero, the lowest temperature possible. I'll address the radiation later. The third factor is the just-discovered existence of possible gravity waves. This discovery is so new and subject to so much doubt that I would stay tuned about it.

The method scientists use to determine relative galactic velocities, and thus expansion rate, is based on the Doppler effect, the same effect we can witness by listening to a passing train whistle. As the train is approaching, the pitch or frequency of the sound is high because the sound waves are compressed by the speed of the train. After it passes and is receding, the frequency is lower because the waves are spread out. In the same way, astronomers look at light waves and use the change in frequency (color in the case of light) to determine velocity in relation to the earth. Looking in any direction, they find that the change is always toward the side of decreased frequency, meaning that the body is moving away from the Earth. They call this the Red Shift. Furthermore, the further away the galaxies, the greater the Red Shift.

So far, it all seems logical. But the Red Shift isn't positive proof of expansion. There is a theory, developed in 1930, that says anything between a star and our telescopes that could affect the frequency of light might confuse the picture, or negate it entirely. This theory, called "tired photons", has been discounted by most cosmologists. But not all, by any means. If space were nothing, it would be difficult to suggest an alternative to the velocity-color relationship. But we've already seen that "our" space is not nothing. It was brought along with matter and energy in the presumed Big Bang. It is said to respond to gravity. It is supposedly full of invisible (or dark) matter and energy. It is called a "fabric". How can we ignore the possibility that in light's billions of years' journey through this mysterious soup, some slight retardation might occur. As an example of light's susceptibility to outside factors, look at the supposedly constant speed of light. That speed is anything but constant when light passes through a medium such as air or water; it can be so attenuated as to approach zero. Why couldn't frequency be affected by something other than relative velocity? If the long, hazardous trip through whatever space is and whatever it contains has even the slightest retarding effect on light's frequency, the observations of the Red Shift could be just what we observe, without any velocity effect at all. The further away the galaxy, the more retardation it would suffer, and voila! The Red Shift. At the time this theory was first rejected, astronomers did not know there were such things as dark matter and dark energy filling the "vacuum" of space. Admittedly, there are weaknesses in this theory, the main one being that images from distant galaxies might be fuzzy if the photons were "tired". But that too is speculation. Furthermore, the fact is the photons from very distant stars *are* fuzzy, as you can see for yourself by studying photos taken by the Hubble space telescope.

Now let's look at the background radiation. For a while, scientists could not understand why the temperature of that radiation is so uniform throughout the universe. Their calculations indicated the matter/energy involved in the beginning of the BB would not have had time to transfer heat from one portion to another, and thus there would have been temperature differences. But then the mathematicians sharpened their pencils and found that there *was* enough time after all. Fine. But now comes the big question. The current temperature of the background radiation is uniformly 2.7 degrees. This is the temperature calculated to have existed at the moment of the BB. So are we expected to believe that the temperature has remained steady throughout the universe for 13.8 billion years? No heating, no cooling? As an engineer, I would say either it did not happen that way, or it's a miracle.

A possible consequence of all this is that the universe was not created by a big explosion at a singularity, followed by expansion of space to its present size. If the creation event occurred all over the universe at once, and the Red Shift is at least partly an artifact, the phenomena we see would make sense. We would no longer be faced with looking past the "end" of the universe, or seeing the Big Bang. The background radiation that is so often quoted as proof of the Big Bang might be the result of what could be called the Bigger Bang; simultaneous creation of the universe in place. Or maybe it's something else entirely, perhaps connected to dark matter or dark energy.

Another possibility is: if the Big Bang did not happen, the universe might be much older than 13.8 billion years, may even infinitely old.

If version 2 of the Big Bang is true, then it no longer describes the creation of anything. It says only that the universe once was in a much smaller volume, and in a much more chaotic state. It does not explain where this massive glob came from or what might have sent it rushing out into the void in the first place. One would presume that gravity should hold it together; that, in fact, it would constitute a single black hole of unimaginable proportions (see chapter 9). And, of course, since some scientists consider a black hole to be a singularity, we are back to version 1.

The same question could be asked about the standard Big Bang theory: what would cause all that mass and energy to explode outward?

You might ask: what is the purpose of this chapter? Well, since we're discussing miracles, we might as well know what miracles we're talking about. If anything, simultaneous creation of the universe is at least as miraculous as a single-point explosion, and a lot more likely. And if it's infinitely old, the miracle is compounded. But no matter how one looks at it, the fact is that scientists have not a clue how the original matter and energy of the universe came to be created.

So far, there have been only two serious hypotheses that scientists have been willing to consider for the creation of the universe: the Big Bang and the Steady State theory. The latter assumes that matter and energy are continuously created and destroyed, one particle at a time. See Chapter 6 for a description of how this miracle has actually been observed on a very small scale. The Steady State theory gets us to the Buddhist philosophy that the universe has always existed, ignoring the miracle of infinite time. The Steady State model also has been widely rejected, even though it was touted by several very prominent scientists, because it ran afoul of expansion and the background microwave radiation. But scientists do not seem to want to look hard at the possibility that the universe was created nearly as it is. They are able to accept one big miracle, the Big Bang, but to face a nearly infinite number of miracles may be too much. Even harder for them (or anyone) to accept is the idea that the universe is infinite, in both space and time. Yet that may be what the data say.

    I will stick my neck out here and make a prediction. I predict the day will come when scientists begin to doubt the Big Bang theory. They will then be faced with an even more complex creation scenario, one that may some day get us to the truth. That is, assuming that the truth of creation is knowable by men, something I personally believe is unlikely. And being unknowable is another definition of a miracle.

5. The Properties of Light

To understand just how strange light is, we first have to consider the nature of waves. And then we must look at water waves in the ocean and sound waves in air.

Essentially, waves are a periodic, traveling undulation of some medium which serves to transmit energy outward from a source. It is characterized by a wave length, which is the distance from one wave crest to the next; an amplitude, which is the height of the crests; and a frequency, which is the number of waves passing any point in a given time. Bear with me here, because there is a purpose to all this.

Now let's take water waves. Most laymen have the idea that waves travel across the ocean as great independently-moving masses of water that eventually throw themselves upon the land. But the fact is, as you can verify for yourself by swimming out past the surf into the ocean or even paying attention in the bathtub, waves do not consist of water moving horizontally at all. Any particular element of water involved in a wave moves mostly up and down, with a little circular motion thrown in. Let's look at how such a wave works:

Something starts it, usually heavy wind. But for simplicity, let's say a small underwater earthquake jars a spot in the ocean and forces some water upward above the point of the earthquake. That elevated position means that energy, called "potential energy", has been added to the water molecules involved. Gravity then pulls the elevated section of water back down. Now it no longer contains excess potential energy. But that energy has to go somewhere, so it is transmitted to the element of water next to the original ring. In other words, the downward motion of the water forces the nearby water upward. Because the earthquake force was directed outward, the transmittal of energy is in the same direction, forming a ring of ever-increasing diameter. This process continues, sometimes for thousands of miles, with the energy of the earthquake travelling horizontally but the water itself going nowhere except up and down. The water is simply the medium through which the energy moves. Only when it reaches a shore where it can no longer move up and down does the wave move forward, finally dissipating its energy by knocking down a few bathers or houses.

Next, let's look at sound waves in air. Again, assume a point source of sound, which injects energy into the surrounding air. In this case, the energy travels outward from that point as a sphere rather than a circle. Since air is compressible, unlike water, it has no need to bounce up and down to transmit the energy. It simply compresses momentarily. Of course, compressed air contains more energy than air at atmospheric pressure. But without any container to hold it, the compressed air at the surface of the sphere cannot stay compressed for long. So it expands, forcing the element of air ahead of it to compress. This process, like the one occurring in the water wave, continues as energy is transmitted from one air molecule to the next in the form of a wave. But the air itself does not move appreciably; it only moves back and forth slightly, serving as the medium through which the wave of energy propagates itself. The intensity of the sound decreases rapidly with distance as the sphere grows larger and the original amount of sound energy is spread over an ever-increasing volume.

And what, you ask, has all this to do with light? Well, light also travels in waves. It exhibits all the characteristics of waves. It comes in an infinite variety of wave lengths, and those wave lengths determine what color it is and even whether it's visible at all. Ultraviolet light has very short wave lengths and infrared very long. Radio waves are the same as light, except with extremely long wave lengths. X-rays are a form of light having very short wave lengths, and gamma rays are shorter still. You can demonstrate to yourself that light is a wave, simply by holding two polarized lenses up to a light, turning one of them 90 degrees, and noting that the light is blacked out. Polarized lenses contain a myriad of minute, parallel "slots" through which light passes; they work by cutting off most of the light waves that are vibrating at an angle to the slots, allowing only those that are parallel to the slots to pass through, thus reducing glare. When two lenses are turned 90 degrees, all the light is eliminated.

So now that we've established how waves work and that light exists as waves, the question facing us is: how does light travel through the nothingness of space? What is the medium it uses to propagate itself? In chapter 2, I mentioned the futile efforts to identify the invisible "ether" that was thought to permeate space in order to explain how light waves travel. According to nearly all scientists, there is no ether. It would appear inexplicable that light waves can propagate through empty space at all (assuming space actually is empty - see chapter 4).

But then came another discovery. Partly because of the paradoxes presented by light and its electromagnetic cousins, it was necessary for scientists to invent another way of thinking about them. That invention was quantum mechanics, which will be discussed further in chapter 20. Suffice to say at this point that it is a mathematical theory based on the assumption that energy, particularly radiant energy such as light, is not emitted smoothly and continuously but in discontinuous pulses or packets. These pulses are called quanta, and their energy is carried by packets called photons. (An interesting sidelight is that four centuries ago Sir Isaac Newton theorized that light was in the form of particles.) Photons are reported to have no mass but significant momentum.

    Now here's the thing about this theory. The fact that light consists of bursts of energy, as if it were made of particles, explains a lot of observations, but does not explain how light also acts like a wave. Some scientists have speculated that light travels in packets of waves, but this does not seem to fit the facts and is rejected by most scientists. There is much more about this phenomenon in chapter 20.

    Momentum is defined as mass times velocity; if photons have zero mass, they should have zero momentum. But they do have momentum because they exert a force against objects on which they impinge. And another thing: if, as mentioned in chapter 2, they have no mass, why do they respond to gravity? Why do they get sucked into black holes? Yet if they did have mass, they would not be able to reach the speed of light because, as we saw in chapter 1, it would require an infinite impulse or force to accelerate them. Photons need no significant impulse at all. Turn on a light and the photons zip out at the speed of light with no mechanical force driving them. As far as anyone knows, they don't accelerate from zero to light speed; they reach top speed instantaneously. Obviously, the electromagnetic spectrum, including light, is very different from anything else we know about.

    But that's far from being the strangest thing about light. Second place in the oddness hall of fame belongs to that fantastic universal constant, the velocity of light. Einstein showed that no matter how you measure that velocity, it is always the same 186,330 miles per second (in a vacuum). Light goes considerably slower when it passes through a medium, because it gets continually absorbed and re-emitted, so let's confine this discussion to a vacuum. Even if the source of the light is moving away from us at, say, half the speed of light, the light is traveling at the usual 186,330 mps, when measured relative to us. No matter how fast or in what direction the earth is moving, light is always moving at exactly the same speed relative to wherever we are. Think about that for a moment. If you were sitting in a spaceship just floating along in empty space and could accurately measure the speed of the light streaming past from behind you, it would be the usual velocity in a vacuum -- scientists call it c. Then say you started your rockets and reached 20,000 miles per hour (relative to the spot in space where you were before). If you now measured the velocity of light streaming past you in the same direction you were headed, it would be c. In other words, it would appear the light had speeded up by 20,000 MPH. But if a person in an idle spaceship watching you go by were to measure the speed of light moving in the same direction, it would still be c to him also.

    What's going on here? That's a good question. Here we have this energy whizzing around without being driven by any known force, at a velocity forbidden to any kind of mass, acting like waves when we look at it as a wave, acting like particles when we look at it that way, moving at an apparently constant velocity relative to everything, having no mass yet capable of pushing a comet's tail out and reacting to gravity. Light and its relatives follow no known system of logic. Einstein's Relativity theory explains it all by invoking the mutability of time and space with velocity, whereby both are decreased as velocity increases, thus presenting us with yet another miracle.

    This fact of relativity brings up a big question, at least in my mind. If a galaxy is moving away us at higher than light speed because of expansion, how does this square with the principle that light travels at c relative to any other body and that nothing can move faster? Is the earth not a body, and is another galaxy not a body also? Can they move faster than light relative to each other? To say it's due to expansion of space seems sophistry to me.

Another question one might ask: what causes light, especially if it's in the form of particles, to rush around at light speed in the first place? What drives it? Again, the answer is: nobody knows. It seems the more questions you ask about light, the fewer satisfactory answers you get.

Another thing to remember is Einstein's amazing equation $e = mc^2$. The velocity of light c is a constant. The equation very accurately describes the relationship of mass to energy, in other words the relationship of everything in the universe to everything else. The velocity of light is like a universal glue that ties us both to the cosmos and to the atoms that make up our own molecules.

The weirdness prize goes to the property of light whereby a photon can be in two places at the same time. This particular craziness will be discussed in the chapter on quantum mechanics (Chapter 20), along with some of the total insanity that arose out of it.

In some ways the behavior of light is one of the most fascinatingly inexplicable of all the phenomena surrounding us. Considering what it does, completely ignoring known laws of logic or physics, one gets an eerie feeling, a sense that somewhere in its mysterious -- no, let's say it -- its miraculous ways may lie a secret, maybe a major secret, of the universe.

6. Strange Behavior of Elementary Particles

In the previous chapters, we saw how scientists seem to have no philosophical problem about the "supernatural" aspects of the Big Bang theory. Somehow the miraculous creation of everything out of nothing gets lost in the complexity and grandeur of the theory, and the subject becomes merely a search for confirmation, and then an answer in itself. But there is another miracle that even atomic physicists call creation. Furthermore, unlike the Big Bang that can only be deduced, it is a phenomenon that can be and has been directly observed. That is the continuous creation and annihilation of matter inside certain elementary particles in the atom. But before we get to that, let's explore some other miracles inside the atom.

As everyone who has taken a chemistry course knows, the atom, already very small, is mostly empty space. There is a tiny nucleus where nearly all the mass is concentrated, and then, far away from the center of the atom, various numbers of electrons whizz about, forming a kind of cloud. In the nucleus we have protons, which are positively charged particles, and neutrons, which are almost identical to protons except for being electrically neutral and very slightly heavier. Neutrons could be considered protons that have absorbed an electron; however, physicists prefer to look at them as consisting of differently charged tiny particles called quarks. At one time not long ago, protons, neutrons, and electrons, the building blocks of atoms, were considered to be the elementary particles. For the science of chemistry, those three are still enough to describe nearly everything one needs to know about atoms.

But then nuclear physicists began looking into the proton, and they discovered some amazing things. For one:

the proton too was mostly empty space. They delved deeper and discovered other particles, such as positrons (similar to an electron except oppositely charged) and neutrinos (also similar to the electron but of neutral charge). They bombarded various nuclei with heavy particles and observed an astonishing number of new particles being emitted, some of them pretty esoteric: muons, pions, quarks, charmed quarks, all sorts of goodies. They thought these were the real elementary particles. They found various forces at work: weak forces and strong forces and electromagnetic forces. They found so many strange and wonderful things that I sometimes had a flashback to my old Nuclear Engineering professor who claimed one day that all those "elementary particles" were so much hogwash; they were being formed in the process of trying to break the nucleus apart to look at it. He said the nucleus was just a little bundle of pure energy. As it happens, it is obvious he was not correct, although a few of those particles did turn out to be just what he predicted -- products of the bombardment and not elementary species. However, based on the advanced state of the art since then and a lot of very persuasive data, it appears that there really are essentially all the particles in there that the physicists claim.

The latest theory is that there are twelve elementary particles: six kinds of quarks, three kinds of electrons, and three kinds of neutrinos. In ordinary matter, there are only four elementary particles: two kinds of quarks, one kind of electron, and one kind of neutrino. (However, when it comes to "ultimates", one should stay tuned.) Current understanding is that both protons and neutrons contain three quarks, a strange energy called gluon,, and very little else. At this point you may well ask: Quark? What the devil is that? I'm glad you asked, because this is one of the many scientific mysteries that pervades our world. In all my researches into your question, the only attempt I've found to an answer is that a quark is a "point-like" particle. Scientists can describe its mass, its electric charge, and its spin, but they have no idea what it actually is. The reason for that is that no one has ever seen one, and the physicists are saying that probably no one ever will. There appears to be some kind of binding force in the proton and neutron that acts more strongly on a quark as it gets farther away from the center of mass. This is, of course, exactly opposite to any other remote-acting force known to science (except possibly "dark energy", whatever that is). Other forces, such as gravity, magnetism, electrical attraction and repulsion, etc., become weaker with distance. But the internal forces in the proton get stronger with distance, almost as if they were connected by rubber bands. The result of this phenomenon is that as the physicist attempts to pull a quark out of the proton to look at it, the invisible forces holding it back increase exponentially as it moves away, until by the time it reaches the outer boundary of the proton, the forces have become too great to overcome. So far, there is no explanation for this. It is illogical (by our system of logic) and follows no known physical law.

But there is another, even stranger fact about neutrons and protons. The mass of a proton is 938 mega-electron volts (MeV). The contribution of the three quarks is only 11 MeV. The rest of the proton's mass comes almost entirely from the gluon, which is a massless energy. Somehow, energy not only can be converted to mass ($E=mc^2$), but it actually *is* mass. Except it has no mass. If your brain is beginning to bubble, you are not alone.

Nor is this all. Another "impossible" phenomenon that scientists are just now beginning to try coping with is the discovery of paired or "entangled" particles that respond to each other's movements no matter how far apart they are. The chapter 20 with which I've been threatening you will go into this further.

This brings us back to the original observation of particle creation. As the scientists peer into the proton looking for quarks and gluon, they also see tiny specks of matter suddenly appearing out of nowhere and then disappearing into nowhere. The scientists find this very interesting, and they continue to investigate it, as they should. But again, we have to ask the big question. According to what natural law does matter appear out of nothing? How does it work? Is it even possible within our range of knowledge or common sense?

What's more, it's not just inside protons that this phenomenon occurs. Scientists are saying that it happens continually in the "emptiness" of space. If true, space is a mighty different kind of nothingness than we can conceive of. But then again, we already know space is a different nothingness, because it was supposedly "created" by the Big Bang. Before that, there was --- nothing. At this point, words begin to lose their meaning.

When it comes to the idea of an "ultimate" particle, we enter into a philosophical quandary. What is an "ultimate" particle made of? For example, if there are six different kinds of quarks (whatever they are), they must be composed of some different components, and the "recipe" for each must comprise different amounts of those components. The latest and by far the most difficult and esoteric theory to explain this mystery is "string theory", which merits its own chapter.

7. String Theory

    This theory (briefly), proposes that everything, electrons, positrons, neutrinos, quarks, gluon, photons gravitons, etc, etc. is composed of miniscule vibrating strings. The strings are either in the form of loops or are open-ended; other forms are being proposed. The type of material or energy created by a particular string is determined by the frequency of its vibration. Just to keep things interesting, these strings exist not in our 4-dimensional world but in a fantastic world that has anywhere from nine to thirteen dimensions.

    Since we've shown that the universe is very different from what we can understand, string theory gives us a glimpse into a possible existence far removed from anything we can visualize. But even if we could accept the idea of eleven dimensions, the theory leads to two more unanswerable questions. One: what are the strings made of? And regardless of what the material might be, what is *that* made of? Two: what causes the strings to vibrate, not only at different frequencies but at *any* frequency? Frequencies tend to degrade or accelerate; if that happened to the strings that created an electron, for example, the electron would change into something else. In my research into this theory, nowhere have I seen reference to a scientist even asking these questions. And when you think about it, you realize there can be no answer to the questions of what any elementary particles are made of, and what makes them move. It's just like the paradox of infinity, or what makes light move. Everything needs to be made of something, yet there appears to be no comprehensible way a real elementary particle can exist, because it will always be composed of something smaller. Eventually we will be left with a "singularity", an infinitely small point. Does that mean everything is made of infinitely small points? Is it composed of black holes? Are we back to the King's dream again?

We seem to be left with several more miracles.

One of the basic items in string theory is a curled-up six-dimensional particle or space called a Calabi-Yau space. One of the conclusions many string theory scientists have come to is that the "fabric" of space can rip apart within these string structures. It quickly repairs itself, luckily; if it did not, presumably the universe would explode or collapse or do something equally unpleasant.

One of the reasons so many mathematicians and physicists are excited by string theory is that by using it they can begin to reconcile gravity with the other forces in nature, something neither relativity nor quantum theory can do.

I will not try to go into all the twists and turns of string theory, partly because it's far too complex and convoluted; how do you visualize eleven dimensions? The other reason is that, unlike all the other phenomena we see around us, the tiny string world, if it exists, has not and cannot be seen or measured in any way. It is entirely buried in a world of mathematics; a world created in the minds and computations of scientists who were looking for a Theory of Everything (TOE). In my reading, nowhere have I seen a reference to how the idea of multiple dimensions originated. Since one of the goals of almost all modern physicists is to reconcile the force of gravity with the other three known forces in nature, I suspect (but do not know) that the mathematics necessary to obtaining a TOE required all those extra dimensions in order to work. Even then, physicists and mathematicians would run into mathematical quandaries as they continued trying to build a world in which everything fit. When it didn't, they invented mathematical solutions to each problem in turn, all without any way to measure the results.

I admire these guys. They are so smart it makes one's head spin. But as an engineer, I have been taught that if a theory is to be accepted, it must be subject to test and to reproduction. String theory does not even make any objective sense in terms of what we think we know of the universe.

There's something else. I know it's presumptuous for an ignorant outsider to question the brainiacs, but there is a piece of string theory that appears to be downright wrong. The little strings are reported to be about the size of something called the Planck Length, which is about 10^{-33} centimeters. That's a decimal point followed by 33 zeros and a one; inconceivably small. Yet the mass of a string is reported by string theorists to be about that of a grain of dust, namely one millionth of a gram. If an electron, for example were made of only one string, its mass would be at least the same millionth of a gram. But the electron's mass actually is about 10^{-33} grams, many billions of trillions of times less than the string it is supposedly made of. This wouldn't be the only peculiar fact about the universe, but if it's true it would seem to fit nicely into the miracle category.

Most of the information in this chapter came from a book called "The Elegant Universe" by the noted physicist Brian Greene. It should be sold with a bottle of aspirin.

8. Gravity

Everybody knows how gravity works: you throw an apple into the air and it comes down and hits Isaac Newton on the head. Most people who paid attention in physics class also know that the force of gravitational attraction between two bodies is directly proportional to the mass of each of the bodies and inversely proportional to the square of the distance between their centers of gravity. The term "center of gravity" is not a misnomer; the earth, for example is trying mightily to pull that apple down into the center of the planet.

Ever since Galileo took a chance that the Leaning Tower of Pisa would not tip over, and he dropped those two unequal iron balls, people have known that in the absence of wind resistance, everything, no matter what its weight, falls to earth at the same rate. That's because even though the force of gravity on a heavy object is greater than on a light one, the force required to accelerate it is also larger by exactly the same amount. What about a feather, you ask; why doesn't it fall as fast as an iron ball? Wind resistance, as mentioned above. As we all learned in grammar school (hopefully), in a vacuum a feather does fall at the same speed.

There are splendid equations describing quantitatively exactly what gravity does. We can calculate just how fast and high a satellite has to travel in order for centrifugal force to exactly offset gravity and keep the satellite in orbit, even a synchronous orbit. We can describe gravity's effects right down to the last decimal point and out to the "edges" of the universe, and know that our basic equations will always be correct (except for some small deviations due to Einstein's dratted Relativity). Sometimes we can't calculate the relative motions of several bodies, such as in the solar system, but that's because the equations get too complicated for us to handle. Even so, our fundamental gravitational model has never been shown to be wrong. It is so reliable that its use has exposed the invisible dark gravitational force holding the universe together, mentioned in chapter 2.

However, there's a funny thing about gravity. No one really knows what it is. Granted, scientists say it's a bending of space, but we've already discussed the circular logic of that rationale. To illustrate just how little scientists really know about gravity, there is still a major debate underway about the speed of gravity. Most mainstream scientists say that gravity acts at the speed of light. They have some support for this, although much of that support seems to be based on the belief that gravity is either a wave or a particle (does that sound familiar?) and that neither of these can exceed "c" in velocity. But they have produced no convincing data proving that. Measuring the speed experimentally is extremely difficult, although it has been recently reported that the speed has actually been measured by observations of a distant pair of pulsars that are circling each other. Not everyone accepts the results of that experiment.

On the other side, various scientists have quoted experimental data that indicate the speed of gravity is anywhere from a modest twenty times c to nearly infinite. The wide range of these measurement results show how difficult such tests are. One of the recent fast-gravity proponents was an astronomer named Dr. Tom Van Flandern, who wrote a technical paper on the subject. Unfortunately, in later life Van Flandern lost what little credibility he had when he went off the rails and claimed that a large rock formation on Mars that looked like a human face at some angles was evidence of an advanced civilization 3.2 million years ago. The popular press loved it.

The analogy most scientists use to try explaining their concept of gravity and the bending of space to the general public contains two ludicrous errors, mistakes that say much about their own conception. They say that space is like a giant sheet of rubber, and that a large mass like the earth is sitting on the sheet, pushing it down. Then when another mass, such as a meteor, approaches, the new mass rolls down the slope and into the earth. The larger the mass on the sheet, the more it stretches the rubber, causing stronger gravity. And the closer the object approaches, the steeper the slope for it to roll down. The acceptance of this idiotic model makes one wonder when the art of thinking died in the scientific community. Ask yourself two questions: one, what would cause the earth to push down on the sheet? Is there some other source of equal gravity under the earth, pulling it? And two; even assuming the sheet was pushed down somehow, what would cause the approaching object to roll down? Again, is there some attractive force under the sheet?

This is another example of the circular reasoning that appears whenever scientists try to explain gravity. Gravity is a bending of space. Space bends because a source of gravity is in it. Pigs can fly, as soon as they get wings.

The fact is that gravity remains a mystery. People have postulated gravity waves, gravitational fields, and gravity particles, but so far no one has been able to measure any of them. A possible exception is gravity waves, which some scientists think they may have observed. And in the unlikely event it turns out that gravity acts faster than light, then all of the above would be disqualified.

Einstein tried to develop a Unified Field theory that would tie all the major forces of the universe together. He succeeded in unifying most of the forces, namely electromagnetism and the strong and weak forces of the nucleus. But he failed to account for gravity. Since then, the only half-way promising lead has been the string theory discussed in Chapter 7, which some scientists hope will lead to a Theory of Everything. But you have already seen how esoterically foreign the string theory is compared to our normal conception of reality. Although it combines gravity with the other known forces, it remains merely an extremely complex mathematical concept. No one has invented a way to test it experimentally, and it is not widely accepted.

So as far as anybody knows, gravity just is. Saying it's a bending of space is merely describing a possibility, it explains nothing, and it makes no logical sense to boot. Saying it's a particle makes little sense. What particle? Calling it a wave does not get us very far, at least at this point in our knowledge. What wave? Through what medium? Saying it's a phenomenon out of an 11-dimensional world gets us to the point of super-miracles.

Now let's think about what would happen if gravity did not exist, or if the attraction occurred through electrical charges or magnetism or some other less neutral and benign force. Of course, if there were no gravity, there would be no universe of stars and planets at all; there would be just a gigantic cloud of gas floating throughout space.

If the force of gravity were exerted by means of opposing electrical charges, the earth could not exist in its present form. It's difficult to imagine exactly what such a world would be like, but one thing is certain: at best it would be a seething mass of energy, a Pandemonium of electrical charges and discharges, a lightning-filled nightmare that would probably tear itself apart before it got started.

If magnetism or something like it were the binding force, the situation would be better but still would be untenable. Every time a wire or any other conductor passed through the powerful magnetic field required to hold the earth together, a strong electrical current would be induced. Every moving metallic object would be highly charged. Life could not exist; electrocution would await any that did. Worse yet, most objects would not respond to the magnetism, and would float off into space.

It is absolutely essential to existence, and certainly to life, that gravity behaves exactly as it does, gently holding us and our pets and briefcases against the surface of the earth but having no other effect on us (unless we are careless enough to fall off a cliff or into the sun), holding our atmosphere in a benign hug to prevent its pouring out into the blackness of space, pulling the huge mass of the sun together so strongly that it generates a perpetual thermonuclear reaction to provide us heat and light. The force of its inexplicable, invisible existence is like a gift -- a gift of calmness, of safety, of light and heat, of life itself. It is nothing less than a miracle.

Of course, gravity is not that benign throughout the universe. For the myriad stars that get pulled into black holes, or that run out of gas and explode, or that get too close to other bodies, gravity can be catastrophic. But we don't live in such a place. As shown in Chapter 10, that fact too might be considered a minor miracle.

To conclude this chapter: one thing seems crystal clear to me. If space can be created along with matter and energy, if it can be curved by gravity, if it can transmit light and other electromagnetic radiations, if it can produce momentary specks of matter, if it contains massive amounts of dark matter and dark energy, then space is a continuous medium of some kind. I don't care if it's called ether or soap suds; it exists. And its existence is a total violation of the known laws of physics, thus a miracle.

9. Black Holes

Einstein formulated his General Theory of Relativity in 1919. It was the first exposition of the idea that time and distance varied with velocity, especially near the speed of light. This theory also predicted (although Einstein didn't know it) that if a star burned out and collapsed, the density of the remaining material would increase to unimaginable values. For example, if a star with three times the mass of our sun were to collapse, its diameter would decrease from 1.25 million miles to only 10 miles. In 1939, J. R. Oppenheimer, working with Einstein's principle, predicted that the gravity density of such a collapsed star would be so immense it would prevent anything, including light itself, to escape. Thus we would have a Black Hole. Ironically, Einstein did not believe it. It seems Einstein had trouble believing many things about modern physics. He is not alone.

But in 1970, a black hole apparently was discovered, and since then a great many others have been found. In fact, it appears that at the center of every spiral galaxy there is a black hole so massive that its gravity helps hold the galaxy together. At the center of our galaxy, the Milky Way, scientists believe there resides a black hole that is 4.2 million times as massive as our sun. There are other galaxies out there whose central black hole contains over a billion solar masses. All the stars in a spiral galaxy are rotating around their enormous central masses, like planets around a star. Of course the black holes can't be seen directly, but their existence can be inferred by the close-in stars that are whizzing around them, as well as the x-rays emanating like a death scream from the unlucky stars and other masses that fall into them.

What has all this to do with miracles? In this case, it has less to do with the facts of black holes (assuming they are facts) and more to do with what various scientists have said about them. One nearly inexplicable factor is the very existence of a black hole in the first place. As shown in Chapter 8, the force of gravity between two objects is directly proportional to the masses of the two objects. But, as discussed in Chapter 5, light has no mass. Therefore, it should not be subject to gravity at all. Yet the very cause of black holes is the inability of light to escape their gravitational force. One must ask: which is it, gentlemen? Does light have mass, or do black holes not exist? Or is something occurring that follows no known natural law?

Actually, some (but not all) scientists say exactly that. Within a black hole, they say, the known laws of physics do not apply. It would be interesting to know why they think they know that. At least they're honest about it, although they would never admit they have defined a miracle.

Some scientists, possibly including those above, say a black hole is a "singularity," that even the super-massive black holes at the center of galaxies have collapsed to an infinitely small point, similar to the wonderful speck that supposedly exploded into the entire universe. If that's not a miracle, what is?

But there's more. Many cosmologists speculate that a black hole may be the entrance to a tunnel between two widely separated points in space, allowing future space travelers to reach stars that are thousands of light years away. Some even think we can reach a parallel universe through one of these "worm holes" (see chapter 20). How a gigantic collection of matter can metamorphose itself into a pinpoint or a tunnel is never entirely clear in their enthusiastic expositions. In my always-humble opinion, these people should report immediately to the nearest asylum before they hurt themselves. But then I'm not a cosmologist.

To their credit, most scientists don't buy the worm hole idea. But it is a testament to academic gullibility that the gentleman who first hallucinated this theory is still receiving our tax money in the form of grants. And the popular press loves him, as do the science fiction writers. It does not seem to occur to these enthusiasts that they are talking about a first-class miracle.

I must make an admission at this point. After thinking about all the strange and inexplicable facets of the universe and the atom, it seems maybe some of these speculations about black holes might be true. If so, on the team of miracles, they would be in the starting lineup.

Speaking of amazing cosmological objects, consider the neutron star. This is what remains after a large star has, for various reasons, collapsed and caused a supernova explosion. The remaining tiny star is composed almost entirely of neutrons packed so tightly together that, as one scientists described it, the density would be the same as if the entire human population of the earth were squeezed into the size of a sugar cube. Why he chose that gruesome analogy is between him and his psychiatrist. In addition, these stars are spinning at incredible speeds up to 716 revolutions per second. Many of them also emit strong electromagnetic pulses from their magnetic poles; these are thus called pulsars. Neutron stars are not necessarily miraculous, because they have not quite reached the density needed to form a black hole, but that sugar cube certainly is pushing the limits of credulity.

10. The Earth's Position

So far, we've talked about the nature of things that are what they are, regardless of whether or not there was any life on earth or anywhere in the universe. We started with the totally esoteric, immaterial concepts of space and time, and progressed to particular, inexplicable phenomena, many of which are necessary for life but which are mostly miracles in themselves without reference to life. But now we will begin to delve into combinations of wonders that impinge directly on life, so much so that their numbers and perfections stagger the mind. We could accept the idea that some of these conditions might fortuitously combine to be of assistance to life, but how can one explain the fact that *everything*, every single physical, chemical, biological, astronomical, and mathematical variable is exactly right to sustain and strengthen life? In this and the next three chapters we will examine a few of these variables, especially those that could easily have gone very wrong. Scientists call these facts "fine tuning", which they are at a loss to explain.

First consider the position of the earth in relation to the sun. We are roughly 93 million miles from the source of nearly all the energy that drives the processes of life. If we were much closer to that immense thermonuclear explosion, the earth would be too hot. For example, Venus is about the size of the earth, but it is only 67 million miles from the sun. Its average temperature is about 870°F. Life cannot exist there. If we were much farther away from the sun, we would be too cold. Our smaller neighbor Mars averages 142 million miles out, and its median temperature is about -81°F. Many people would like to think that life exists on Mars, but wanting such a thing is not the same as having it. No one has ever found native life there, and for reasons more thoroughly explained in chapter 14, it is doubtful that anyone ever will.

Well, you say, so what? Some planet was likely to exist between Venus and Mars, and it happens to be Earth. Taken by itself, that's a perfectly valid point, except for one fact. If the earth's axis were not strangely tilted 23.5 degrees, and if it did not wobble back and forth from the tropic of Cancer to the tropic of Capricorn every year, and if its rotational speed were not 24 hours per revolution, conditions on earth would be drastically different and probably unlivable, no matter how far from the sun we were.

For example, if the earth rotated around its axis at the same rate as its yearly rotation around the sun so that the same side always faced the sun (like our moon faces us), that side would be fatally hot and the dark side would be fatally cold. Only the thin twilight zone around the edges might have a tolerable temperature. But unfortunately the blazing heat on the sunlit side would long ago have driven the atmosphere and the water out into space.

If a pole faced the sun (like Uranus), the situation would be the same as the above disaster, no matter what the rotational speed of the earth was.

If the earth took, say, a week to rotate around its axis, the temperature extremes would be horrible. Even with our current speed, summer temperatures often rise 25°F in just 7 or 8 hours, then drop a similar amount in the dead of night. If that 8 hours of heating time were expanded to 56 hours, with another 56 hours of cooling due to exposure to the ultimate cold of the night sky, it is likely that the tropics and the temperate zones (where most people live) would experience both days and nights that were impossible for most life. Furthermore, the huge temperature differences would create massive winds and sea currents that would turn the earth into a gigantic Pandemonium.

In a less drastic scenario, if the earth rotated relative to the sun just as it does now, but the earth were aligned with its poles in a plane perpendicular to the sun (a logical position) so that the sun was always directly overhead at the equator, tropical temperatures would be higher and polar temperatures lower than they are now, and the result would probably be much more extreme ocean currents and more severe weather patterns. In fact, if the earth maintained any reasonably constant position at all in relation to the sun, the resulting extreme variations in temperatures would exaggerate weather patterns all over the world, to the detriment of life everywhere. Furthermore, there would be no seasons anywhere on earth. It's difficult to say what the effect of this latter result would be, but at best it would be boring.

Recently, astronomers have determined that many stars are orbited by planets. Most of them are either gas giants like Jupiter or Saturn, or are far too close to their star, or are in an exaggerated elliptical orbit that carries it too far out and then too close. They have found very few planets of the right size and in the right location for life, but they remain hopeful. Even when they do, there are numerous other factors involved in finding another Earth (see chapter 11).

Another even more sinister, though less probable, possibility exists. Suppose our star, the sun, were located differently within the Milky Way galaxy, much closer to the galactic center. That center area is extremely crowded, and a great deal of wild activity is occurring there. As mentioned in Chapter 9, a super-massive black hole, with a mass over four million times that of the Sun, lurks at the very center of the galaxy. Beyond that, the proximity of stars is much greater than it is out closer to the galaxy's edge where we happily reside. In the center, dust clouds are busily creating new stars, and old stars are expiring, often taking other stars with them. Black holes suck matter, including other stars, into their pitiless maws, producing gigantic x-ray bursts. There are stars whirling around the central black hole at up to 20 million miles an hour, some of them so fast they end up being hurled completely out of the galaxy. Radiation levels, including x-rays, are fatally high in many areas. The existence of life as we know it is almost certainly impossible.

But none of these things happen, because the earth *does* rotate in 24 hours, and it *is* tilted, and it *does* wobble, and it *is* 93 million miles from a nicely-situated sun. What an astonishing series of coincidences. Alone, these facts do not constitute a miracle. But that's only the beginning.

For life, especially human and animal life, to exist, much more is required than just the right weather. The next chapter will look at some of these other factors.

11. Composition of the Atmosphere and Earth's Crust

Let's say we just got lucky, and the earth fortuitously happened to locate itself the right distance from the sun, and to have just the right orbital speed, and to rotate just right, and lean correctly, and wobble exactly as it should. Is all that enough to explain why the earth supports life so nicely? Of course not. We could easily have practically no atmosphere, like Mars, or a carbon dioxide/sulfuric acid atmosphere, like Venus, or even one of those gruesome ammonia and methane atmospheres like the larger planets. Or if we avoided those fates we could have a suffocating all-nitrogen atmosphere, or maybe a fiery, all-oxygen one. We could have almost anything. But only one atmospheric composition fits the needs of life as we know it, one composition that allows us all to breathe easily without burning us up, either through accelerated metabolism or by actual flames. That composition is 21% oxygen and 79% inert gases, mostly nitrogen. And that, of course, happens to be the composition we have around us. Some coincidence.

But since we breathe in oxygen and convert it to carbon dioxide and water, that atmosphere would not last long in the absence of other "coincidental" factors. We need something to recover the oxygen that got tied up in our products of respiration, and recycle it back to the atmosphere, so we can breathe it again. And it has to accomplish this at exactly the rate at which we use the oxygen, so as to maintain a constant oxygen content. All of this, of course, is precisely what plants do.

One might ask how the plant world manages to vary its output of oxygen to match the use rate. The oxygen content of the atmosphere has not changed since the days when the human population was a small fraction of its current value. Part of that may be due to the animal population being higher then than it is now. But even so, it is true that no matter how fast oxygen is used, the composition of the atmosphere remains the same. If a huge forest conflagration occurs, or an Industrial Revolution comes along and begins emitting vast amounts of carbon dioxide, or the Iraqis torch 500 oil wells in Kuwait, the carbon dioxide level rises only very slightly, and the oxygen is still there. Why? The answer is that plants grow faster as the carbon dioxide level increases; thus they use CO_2 faster and emit more oxygen. The result is a natural balance, a very stable equilibrium that keeps and has kept the atmosphere on an even keel through thick and thin. Another lucky break, what?

And another thing. Plants need nitrogen for their unique metabolism to work (see chapter 21). We could easily have some inert gas like argon to dilute the oxygen, but then plant life could not grow because argon doesn't react with anything. So we have, "luckily" -- nitrogen, which reacts nicely with plant life but very little with animal life. Just another nice coincidence.

Let us speak of ozone. As people who worry about ozone say: if we had only oxygen and nitrogen in the atmosphere, we would be bombarded with cosmic rays and other evil emanations from the sun, and would get a nasty sunburn or worse. And in fact they're right. We get a good dose of radiation from the sun anyway but, in yet another stroke of "luck", our bodies are built to tolerate it easily. But without some kind of screen to reduce the level of cosmic rays, we might be in considerable trouble. So what does our good fairy, or Lady Luck, or the breakdown of the laws of chance do? It provides us with a self-renewing layer of ozone near the top of the atmosphere, a layer that filters out most of the cosmic radiation without affecting the visible light we and the plant world need.

You'll notice that I said "self-renewing". Ozone is an ionized form of oxygen, with three oxygen atoms instead of the usual two that are in the oxygen molecules we breathe. The ozone in the upper atmosphere is formed by the action of various energetic cosmic rays on the ordinary oxygen molecules in the thin, upper reaches of the atmosphere. Ozone is not a very stable molecule and it reacts readily with almost anything. Various gases that ascend into the ionosphere from the earth's surface, gases such as chlorine from the breakdown of salt in the ocean, methane from marshes and animal flatulence, sulfur compounds from volcanoes, all react with ozone. These processes have been going on since long before man arrived on the scene. Yet the ozone layer remains. It varies in height and thickness in response to various natural conditions, but it is there, doing its job. The reason is, just as in the formation of oxygen by plants, there is a natural, stable equilibrium, in this case between oxygen and cosmic rays on one side and ozone on the other. When something reacts with a portion of the ozone, more ozone very soon forms to replace it. Sometimes natural processes, such as sunspots, temporarily increase the amount of radiation reaching the earth, and then the ozone layer thickens for a time. At other times, it thins briefly, due to various interactions with the complex soup of chemicals and electromagnetic forces working in the atmosphere. With the exception of a few weeks every August when the ozone concentration drops somewhat over the south pole, it is always there.

This equilibrium is the reason why the popular hysteria about mankind's effect on the ozone layer was so misguided. The amounts of naturally-occurring reactants destroying ozone molecules is vastly greater, by orders of magnitude, than all the man-made materials that worry people so much. If the ozone layer can survive the great volcanoes like Pinatubo and Krakatau, which filled the entire atmosphere of the earth with ozone-killing chemicals, it certainly isn't going to notice the puny trickle of "Freon" from hair sprays and air conditioner leaks. Even modest volcanoes like Mt. St. Helen completely overshadowed anything man could emit. The stability of the ozone layer is just another in the amazing series of coincidences that we're blessed with. How many coincidences like this does it take before it becomes a miracle?

But why stop here? The atmosphere is all very well, but having a stable, self-regulating oxygen, nitrogen, and ozone supply still is not enough. We need a hospitable surface over the seething core of molten rock and metal that comprises nearly all of the earth's mass. If we were to sit down and decide what the surface or crust of the earth would have to be like to support human and animal life, we would specify at least the following:

A material of low thermal conductivity, to insulate us from the intense heat just beneath our feet.

A material porous enough to allow water to penetrate sufficiently to nourish plants, but impervious enough to hold surface water in the oceans, the lakes, and the rivers.

A material compact enough to walk on without sinking in dust, but loose enough to till and plow.

A material containing plenty of the right kinds and the right amounts of minerals: potassium, magnesium, calcium, iron, aluminum, etc., to nourish plants.

A material very low in poisons such as arsenic, selenium, or uranium.

We have, of course, just described the surface of the actual crust.

Any one or two of these wonders might be taken as a lucky break. The odds against all of them occurring at the same place, especially when combined with all the factors described in chapter 10, are very high.

And that's far from the end of it. In the next chapter, we'll look at one of the most common substances in the world, and explore its miraculous properties.

12. Anomalous Properties of Water

I can hear you saying: "Now he's gone too far. Water? Water is water. How can it be so special?"

In case you are under the impression that water is nothing to get excited about, let me disabuse you. Water is not "just" water. In fact, I will say it right up front: water is a near-miraculous substance. This chapter will explain why.

In a way, water is a little bit like gravity, something mild and benign that we take for granted but without which life would be impossible. Yet from a comparative viewpoint, water has no business behaving as it does. Unlike the case with gravity, there are good scientific explanations for its behavior, including hydrogen bonding, crystal structures, etc., but that does not change the fact that it is without question the most astonishing substance on earth. Let's look at its properties and compare them with other materials.

First, there is its boiling point. Water is a very small and light molecule, one of the smallest compounds in existence. It has a molecular weight of 18. Yet it is a liquid with a very high boiling point of 212°F (100°C). Every other compound with a molecular weight in that low range is a gas at room temperature and is still a gas far below that temperature. To see how unusual and unexpected water's boiling point is, look carefully at this table of various small molecules, listed in order of molecular weight.

Molecule	Molecular Wt.	Boiling Point (°F)
Methane	16	-257
Ammonia	17	- 28
Water	**18**	**+212**
Nitrogen	28	-320

Ethane	30	-127.5
Oxygen	32	-297.5
Hydrogen sulfide	34.1	-75.3
Propane	44.1	-43.6
Pentane	72	97
Hydrogen telluride	129.6	28.8

The list continues. The numbers jump around, but there are no light molecules even remotely like water in its ability to stay liquid at elevated temperatures as if it had a very much higher molecular weight. If that were not so, there would be little or no water on this earth. Even if the boiling point were, say, 85°F, it would be boiling half the time. Even when it wasn't boiling, it would evaporate so fast that no body of water could survive for long. In fact, the water in our bodies would boil away due to our own body heat.

And then there's ice. Essentially every other liquid shrinks when it freezes, thus becoming denser and sinking. If water did that, the floating ice that forms the Arctic and some of the Antarctic would be at the bottom of the ocean, allowing more water to freeze until the entire northern ocean and much of the southern would be solid ice. The ice would migrate underwater far to the south or north and drastically cool all the oceans and the world around them. The ocean currents that regulate our weather would be entirely different, so much so that life probably would be impossible. Lakes, especially deep lakes, would be ice-cold most or all of the year. So what does water do? It expands when it freezes, floats, and protects the rest of the Polar oceans from freezing. It also allows lakes to warm up in the summer.

Another unique feature is water's heat capacity. Of all the liquids known, only a tiny handful including ammonia and liquid lithium can match the one calorie of heat it takes to raise a gram of liquid water 1°C. The others are not even close; most small molecules hold only about 0.3 to 0.5 calories per gram per °C. Again, this property is absolutely essential for the stable temperature control the ocean provides us.

Take the heat of vaporization. Water requires 540 calories of heat energy per gram to vaporize it. No other volatile material comes anywhere near that figure; most have heats of vaporization well below 100 calories per gram. In fact, the only compounds of any kind that have higher heats of vaporization are some salts of various metals; ordinary table salt is one of these. Why is that important? For one thing, this property greatly slows down the evaporation of water, allowing lakes and streams and even the ocean to exist, allowing crops to be irrigated. For another, the large amount of heat in the water vapor around us is released when the vapor condenses into a cloud, and that energy helps drive the processes that create heavy rain. (Sometimes, of course, it creates a hurricane.)

Look at surface tension. There is no other liquid on earth that even approaches water in its surface tension of 75 dynes per centimeter. Most materials have about one third of that figure. What does this mean? It means that water is much slower than any other liquid to soak into our skins. Part of the wonder of our famed waterproof skin lies in the nature of water itself.

And there is still more. Every other liquid molecule is toxic if not downright poisonous to animal life. Water is the most non-toxic substance there is, not to mention its critical role in all life. If we had gills, we could even breathe it.

How did this amazing substance come to behave as it does? How is it that every aspect of its existence is anomalous and just right? If one or two of its properties were unusual or unique, one could just accept that gift gratefully. But for practically every one of its physical characteristics to be far outside the expected bounds is awe-inspiring. Why do those hydrogen bonds form and hold the molecules together so nicely, when many other molecules that also contain pendent or reactive hydrogen atoms do not? Why, in that same simple water molecule, do those unique crystal structures form that cause it to expand when it freezes? Why does that same molecule hold so much heat, both in the liquid and the gaseous form? If you were to design the perfect fluid to support life, apparently violating the norms of physics in the process, you would design water.

Since most of water's properties can be explained scientifically on the basis of various known phenomena, water is not in the first category of miracle. But the reality of water, in which almost every single characteristic is unique and unexpected, making it exactly right to sustain life and the world that life exists in, seems too perfect to be a coincidence.

13. Unique Constants

We have seen in the previous chapters how everything in nature seems to be just right, as if it had been carefully designed to be hospitable to life. Yet all those amazing features: water, gravity, the atmosphere's composition, the position and motions of the earth, etc., only scratch the surface of the incredible "coincidences" that go to make the universe work, and especially what makes the earth the unique and wonderful place it is.

Everything around us, from the smallest elemental particle to the largest galaxy, obeys a myriad of universal laws. We talk about the laws of physics, the laws of chemistry, the laws of biology, etc., but in fact these distinctions are entirely artificial, invented to help us understand and remember at least part of what we observe in the universe by compartmentalizing our knowledge. In nature, there is no physics or chemistry or biology; everything is part of a seamless whole. The ancient philosopher/scientists understood this better than we do; someone like DaVinci was a multidisciplinary scientist, an inventor, an artist, a military expert, a philosopher. He would never understand the specialization we indulge in today. Even now, anyone who aspires to scientific greatness must be interdisciplinary. Otherwise, he will be leaving out most of the data and observations that impinge on his field.

The "laws" I mentioned apply to all aspects of science and of the universe. We don't know what many of those laws are, but we are reasonably sure they exist and apply to everything. And for every law of nature, there is a mathematical way of expressing it. There are invariably equations that relate some physical phenomenon to others, often to causes but at other times to variables that respond to the same stimuli. And almost always an equation that relates one variable to another contains at least one constant. A simple example is the relationship between the energy of radiation and its frequency: $E = h\nu$, where E is the energy content of a radiant source, ν is the frequency of that source's emission, and h is a universal constant (fudge factor) called Planck's constant. This is a very minor example; many relationships are much more complicated, with multiple constants. But the point is that all these constants exist.

So what, you say? The reason all this is important is that if any of the dozens of constants governing the universe had a different value, even slightly different, the world would not exist as it is. Possibly it would not exist at all. Plank's constant is 6.625×10^{-27} erg sec. If it were 7×10^{-27} erg sec, the universe would be out of kilter. Energies would not balance. Equilibrium, which is practically the definition of stability, would be impossible. The alternative to stability is chaos. The only way that equilibrium could be regained would be for all the other constants to be different as well, and by just the correct amount. However, in that case, the universe would be different also.

In addition to mathematical constants, there are endless physical coincidences that are necessary for our existence. For example, (1) Tim Folger in Scientific American named the following:

a) In the sun's thermonuclear reaction, about 0.007% of the sun's mass becomes energy. If it were 0.006%, there would be no fusion and no sun. If it were 0.008%, there would be no hydrogen left, and the sun would go out.

b) If matter had been more evenly distributed, there would have been no clumping and no stars. If there were less matter, it would all pass into a black hole.

c) If the nuclear strong force (one of the four basic forces of nature) were slightly larger, there would be no hydrogen.

(2) Andrei Ling, a physicist at Stanford, Univ. named the following, among others:

a) If the proton were 0.2% heavier, it would be unstable and would decay. Without protons, there would be no atoms, and with no atoms there would be no universe.

b) If gravity were slightly stronger, stars would burn up in a few million years, leaving no time for galaxies or cool planets to form.

c) If the mass of the electron or the strength of the proton-electron interaction were slightly different, there could be no life.

But the constants and the relationships are what they are, and as a result the world is exactly and uniquely right. Scientists even call it "fine tuning". Is this merely a coincidence? Mathematically, the chances against all these constants and relationships being exactly right are astronomical. In fact, there is no meaningful number with which to express the odds. It cannot be a coincidence. And that leaves -- a miracle.

Some scientists are so overwhelmed by this apparent coincidence that in their desperation to explain it they have invented something called the Anthropic Principle. One rather sarcastic definition of the Anthropic philosophy is "if things were different, things would be different". The Anthropic Principle has been bent into many shapes, one of which is Intelligent Design. But most scientists, being prejudiced against anything smacking of what they consider religious, reject that reading and instead develop remarkably abstruse and unlikely forms of the Principle. One states that there can be no universe without an intelligence to observe it. Here we have a modern resurrection of the medieval belief that it's all the King's dream. Another says, as far as I can understand it, that fine tuning is inevitable because we are here and we could not be here unless everything were fine tuned. That is certainly true, although how it explains anything is beyond me. Yet another proposes that humans adapted by Darwinian evolution to the universe as it is, ignoring the fact that without fine tuning there would be no world to evolve into. A major form states, briefly, that because there is an infinite number of parallel universes, one of them is bound to have all the right factors and constants. And we happen to be living in that one.

    This last proposal, commonly called multiverse, is the only one that seems to make sense to many scientists. Some suggest that other universes sprang out of the same event that created our universe. Others, hypothesizing that the matter in a black hole has been crushed to an infinitely small point, just like the one that supposedly created our universe, suggest the other universes come out of black holes. The implication is that as soon as matter and energy get squashed to a small enough size, they suddenly become capable of creating a whole new universe. Right. Do these guys ever think what they're saying?

    The quandary that fine tuning creates for the scientific community is best summarized by Bernard Carr, a cosmologist at Queen Mary University in London: "*If you don't want God, you'd better have a multiverse.*"

    George F. B. Ellis wrote: "*Even if the multiverse exists, it leaves the deep mysteries of nature unexplained.*"

    Freeman Dyson, a renowned physicist at the Institute for Advanced Studies at Princeton, may have inadvertently given the best description of the problem for most scientists: "*The Anthropic Principle means the universe knew we were coming.*"

Just to complete the litany of amusing ideas invented to avoid admitting a miracle, here are some of the silliest:

1) Life elsewhere is based on something other than carbon (silicon perhaps? Or sulfur?) (Victor Stenger)

2) The whole fine-tuning business is "*subjective anthropomorphism applied to natural physical constants.*" (This substitutes sneers for logical thought.)

3) Inflationary cosmology. Because of something weird that happened in the first 10^{-33} seconds of the Big Bang: *the overall shape of the universe at 14 billion years is much less sensitive to initial parameters than the standard big bang model, and thus the fine-tuning issue disappears.*"
Oh, right, that makes sense.

4) The famous Stephen Hawking proposed that fine-tuning is inevitable because "*the current Universe "selects" only those past histories that led to the present conditions*"
Stephen Hawking is very brilliant. He is also one of the stupidest brilliant people on the planet. See chapter 3 for more of his nonsense.

5) Our universe was created by superior beings from another universe. (Cosmologist Alan Guth, PhD.)

I rest my case.

14. Existence of Life

In the last few chapters, we've looked at an amazing array of so-called coincidences involving the makeup, position, and behavior of the inanimate world, so many coincidences without a single negative that they must be considered beyond the realm of pure chance. But now we come to a mystery that matches the creation of matter and energy in its importance and its inexplicable existence. That mystery is the beginning of life.

Probably on no other subject has there been so much confusion and acrimony as on the subject of how life began. Part of the problem stems from the definition of a single word: evolution. Almost universally, people, including many scientists, tend to defend or attack evolution as if it were a monolithic *thing*, some solitary concept that is either true in all respects or false in all respects. The fact is that there are at least four major meanings to the word.

One refers to micro-evolution, in which changes occur to species in response to environmental conditions. Essentially everyone agrees that this kind of evolution occurs and is continuing to occur (although some of the more noted examples, such as the black and white moths in England, have been shown to be frauds). There is plenty of observable evidence that mutations occur, that not all (just nearly all) mutations are harmful, that certain mutated features may occasionally find themselves positively useful and eventually favored in some changed environment, and that species can and do die out when they are no longer compatible with their surroundings. Fossil evidence clearly shows changes with time in various animals, but always within the same kind.

The second type of evolution arouses the most controversy. That is the Darwinian extension of micro-evolution to macro-evolution, meaning the creation of new species due to the same environmental pressures that cause changes within a species. This difference may seem minor, but it leads to such concepts as the evolution of not only species but also of whole genera and families. It implies the metamorphosis of amoebas into fish, fish into lizards, lizards into mammals. Except for a recently-discovered cross between a monkey and a lemur (both primates), there is no evidence in the fossil record or anywhere else that this theory is correct. The fossil record clearly shows the progression of the horse from the tiny eohippus to Secretariat, but there is no fossil showing where eohippus came from. On the other hand it is impossible to prove a negative, so I expect we are doomed to battle this one out forever.

The third type of evolution involves the development by Darwinian natural selection of organs such as the eye and the ear. This weak theory merits a review of its own, and is discussed in chapter 19.

But the fourth type of evolution is one that is, frankly, impossible. It is the theory that says life arose spontaneously from naturally occurring, inanimate molecules. Not only is such a thing illogical and contrary to common sense, but it can be proven false mathematically. Let's address logic first.

The theory of spontaneous inanimate vivification gained its current popularity during a time when there was a new knowledge of electricity, and at the same time a near-total ignorance of microscopic biology. In Darwin's day and before, men were entirely unaware of the nature of the single cell. They thought living matter was made up of tiny lumps of "protoplasm", an amorphous material containing both organic and inorganic substances. At the same time, throughout the entire civilized world, scientists and laymen alike were caught up in enthusiasm about the new science of electricity. They thought electricity was the missing key to life. For example, they found that an electrical charge properly applied to a dead frog could make the frog's legs kick. Many people thought, in fact, that electricity itself *was* the life force. Between the simplicity of their "protoplasm" and the power of their electricity, it was easy for them to visualize a small, stagnant lake, a primordial soup, and a bolt of lightning transmitting the force of life to that first living cell. They were sure that mankind was on the verge of creating life in the laboratory. It seemed so simple that Mary Shelly could say in her book "Frankenstein", without being laughed out of society, that a scientist had passed electricity through some vermicelli and brought it to life. She may have been referring to a dead worm, but possibly it was old pasta.

    Then came improvements in the microscope, and the realization that just the nucleus of the single cell, the simplest form of viable life, is so incredibly complicated that it still has not been fully defined. The cell is, in fact, almost a microcosm of the entire animal or plant it comprises. It contains mechanisms that can carry out thousands of complex reactions, with attendant transfers of mass and energy. It coordinates its activities through sophisticated electrical and chemical communications among its parts. It contains 60,000 protein molecules, of 100 different compositions. It has DNA, in fact the same DNA that the animal possesses. It automatically keeps all its functions under control and in balance. Its design is far beyond the ability of our most accomplished scientists and engineers even to approach.

    You would think that once this fantastic complexity had been discovered, scientists would revise their belief in the primordial soup and the lightning bolt. For the most part, you would be wrong. This is still by far the most widely accepted explanation for the beginning of life on earth. It is taught all the way from elementary schools to prestigious graduate schools as if it were a fact. Yet, unlike the theories surrounding "evolution" of the second sort, it is not even a valid theory. It is a mathematical impossibility. Here's why.

The men who first formulated the theory of spontaneous life were laboring under two other major misconceptions, in addition to the protoplasm delusion. First, they believed the earth had existed forever, not for just 4.5 billion years. Second, they believed that given enough time, any sequence of natural events that could be conceived, no matter how unlikely, would eventually occur. It was the old story of the monkeys and the typewriters; supposedly, if you put some monkeys in a room, each with a typewriter, and let them type randomly for long enough, they would eventually type all the works of Shakespeare, as well as everything else ever written. That scenario has been expanded to include an infinite number of monkeys working for infinite time. I remember being taught that little mind teaser as a child. It's still being taught, and is believed by most people, including many scientists. Of course the monkey metaphor is a poor one because a monkey would tend to pound on only a few keys. So let's just say we have a random letter generator.

The trouble is, in addition to the fact that 4.5 billion years is a long way from forever, there are events so unlikely they do not and will not happen, no matter how long you wait. Here's an example, using standard methods of statistics. If our monkey (or random letter generator) wishes to type the word "tender," the chances that he will type the letter "t" are one in 26. The chance he will type "e" are also one in 26. But to find the chances that he will type both together you multiply the probabilities. Thus the probability of typing "te" is one in 26 x 26 or one in 676. To type "tender", the chances are one in 26^6 or one in over 3 billion. For 20 letters to produce a reasonable result, the odds are one in (2×10^{28}). That's two followed by 28 zeros. "Hamlet" contains over 130,000 letters. Obviously, the odds against random letters producing sensible results grows very quickly to absurdity. As Kittel and Kroemer say: "Even if the observable universe were filled with monkeys the size of atoms typing from now until the heat death of the universe, their total probability to produce a single instance of *Hamlet* would still be many orders of magnitude less than one in $10^{183,800}$." To show you how far beyond any semblance of reason that number is, several competent scientists have calculated that the total number of elementary particles in the entire known universe is "only" 10^{80}.

Obviously, the number of useful events that could be generated by random chance is extremely limited. The ultimate mathematical limit can be calculated by starting with that number of elementary particles in the universe: 10^{80}. Multiply that by 10^{45}, the maximum number of "events" each of those particles could undergo in each second. Multiply that by 5×10^{17}, the total number of seconds in the 13.6 billion-year history of the universe. Finally, multiply by a billion, just to be conservative. The result is about 10^{150}. That's the absolute maximum number of atomic events that could have occurred throughout the entire universe in all its history. That number, 10^{150}, is probably as high a number as could have any meaning at all.

Now compare that number with the probability of typing Hamlet by random chance (one in $10^{183,800}$.) Then think what would be required to evolve a strand of DNA, which has 204 billion atoms (Michael Graham Richard).

Continuing along this line (although surely you can see where it's headed), what would it take to evolve a single cell? Well, first you'd have to evolve all 60,000 protein molecules of 100 types. The odds against that are $10^{4,478,146}$ to one. You read it right. That's a lot of zeros. The number is entirely meaningless, because it vastly exceeds the total number of anything and everything in the universe, no matter how small.

And that's not all, by any means. Consider that all of those 60,000 proteins would have to evolve at the same general time, and they would have to accumulate in the same spot, all lined up in just the right sequence waiting for the magic lightning bolt to zap them to life. Only that still would not be enough. Some of that double helix DNA would also have to evolve and be hanging around with the proteins. In fact, as Mastropaolo puts it: the water would have to be thick "with polysaccharides, lipids, amino acids, alpha helixes, polypeptide chains, assembled quaternary protein sub-units, and nucleotides, all poised to self-combine into functional cellular structures, energy systems, long-chain proteins, and nucleic acids. Then during an electrical storm, just the right mix of DNA, mRNA, cell membranes, and enzymes are envisioned in the right place at the right time, and the first cell is thunderbolted together and springs to life". When you realize what would be involved in creating the first living cell by random chance, it is incredible that any scientist still capable of dressing himself could actually believe in such a thing. As Humpty Dumpty said in Alice in Wonderland: "Sometimes I believe in six impossible things before breakfast." But that was fiction. Then again, so is the random evolution of life.

As an amusing sideline on this subject, the legendary scientist Fred Hoyle wisely rejected the Darwinist theory of life's formation and realized the impossibility of a living cell assembling itself on earth. He wrote: "*The chance of obtaining even a single functioning protein by chance combination of amino acids is the same as a star system of blind men solving Rubic's cube simultaneously.*"

So what was his explanation? Did he attribute it to a miraculous design? Of course not. He announced his theory that life had come to earth from outer space, possibly in the form of spores. Ho, hum, problem solved. Let those proteins assemble themselves somewhere else; then we don't have to worry about the origin of life, and can go on with our other theories, the ones that can't be disproved mathematically. What's more, an increasing number of scientists, also disturbed by the odds against random creation of life, are also proposing Hoyle's solution. Maybe life came in a comet. They look for life in meteorites, and on Mars. If they can just find life somewhere else, their worries will be over. It doesn't seem to occur to them that transferring the problem to another world doesn't change anything.

Scientific American magazine recently printed a short article by one of their staff writers in which he shockingly stated that the idea of life originating in a swamp somewhere is extremely unlikely. I was just about to send him a congratulatory letter celebrating his breakaway from orthodoxy, when he ruined it all in the last paragraph by speculating life might have formed instead underwater in one of those sulfurous hot water vents associated with volcanoes. Guess he wants to keep his job.

There is a concept called Fermi's paradox. Enrico Fermi, a celebrated nuclear physicist who built the first atomic pile, used an equation devised by Frank Drake to calculate what he considered the odds of other intelligent life existing elsewhere in our galaxy. Like other scientists, he believed that any planet of the right size and the right distance from its star might have evolved life, and a certain percentage of that would be intelligent life. His result was indeterminate but he and other scientists thought it could be a very high number. Then he asked the big question: where is everybody? With so many worlds full of intelligent life, we should have heard from them by now. For decades, governments and foundations have spent billions of our dollars listening for signals, any signals, to indicate someone else was out there. All they hear is static. At one point there was great excitement because they heard a rhythmic signal. But that turned out to be coming from a rapidly rotating pulsar (see chapter 9). It never seems to occur to all these hopeful folks that after the miraculous creation of life on this unique planet that was miraculously created to accommodate that life, the chance of it happening elsewhere may well be zero.

    I hate to be so sarcastic about some of the finest and smartest people in the world, but until they start facing facts and admitting that something miraculous has occurred, they will remain objects of amusement. I'm happy to say that increasing numbers of prominent scientists are coming to the conclusion that life was miraculously designed, and are brave enough to say so in public (see chapter 20). They face severe pressure from the academic thought police to repent, but they have the comfort of knowing that some day the truth will win out. Especially a truth as obvious and as easy to prove as this one.

15. The Minolta Paradox

There is a wonderful parable that pretty well sums up the logical (rather than mathematical) argument against the natural evolution of life. A spacecraft takes off for Mars with an astronaut inside. The spacecraft lands safely, and the man emerges. He is the first man ever to set foot on Mars. He walks a hundred feet over the rocky, orange surface, and then he sees it. Sitting on a rock is a Minolta camera, full of film, batteries charged, fully operational. Does he then say: "Gosh, over these millions of years that camera must have assembled itself by random chance"? Of course not. But when we find an infinitely more complex mechanism called a human being, we say exactly that. Why?

The Minolta Paradox is not entirely fair, I suppose, because the camera is inanimate while the human is alive, with the dynamic force of life driving every cell in his body. But we've already seen that the life force could not have evolved by random chance. Furthermore, the complexity and perfection of living creatures argues logically against totally random assembly even after the problem of life's beginning has been either accepted, solved, or ignored. We saw how complex a single cell is, with its 60,000 proteins, its DNA, its communications and control systems, etc. The operation of an animal, especially a human being, is so much more complex than a single cell that the greatest scientists in the world, after centuries of research, have only begun to understand some of its workings.

To delve deeper into this subject, let's look at DNA. All educated people know at least a little about the double helix, and how it carries the multitude of genes that define our individual self. The gene code is in four nucleic acids attached in one of a nearly infinite number of combinations. DNA can replicate itself by unwinding into two single helixes, each of which can then react with the correct amino acids to build, step by step, an identical twin wound into its own structure to form the double helix of DNA. But now there are two DNA's.

This DNA has mind-blowing powers. Among its innumerable jobs, it provides instructions for forming millions of different proteins, miraculous molecules some of which are so complex they seem almost intelligent (see chapter 19).

The news media were full of stories about scientists cracking the code and being able to describe the entire structure of DNA, and what genes are where. This effort took decades of arduous work by hundreds of researchers, a testimony to the extreme complexity of the molecule and the entire system it directs. Without a doubt this is a fine achievement. It may well lead to improved medical procedures. But now let's ask a few questions. How does the organism read the code? How is it determined for the developing fetus that this particular amino acid in this position plus that amino acid in that position determines the shape of its eventual nose?

Even more to the point, how did such a complex molecule ever develop in the first place? Of what evolutionary advantage is a code with one or two nucleic acid groups stuck on a short chain of carbon atoms? Yet that's the best that one could hope for as a first step in "evolving" a meaningful DNA helix by random chance. As described in the previous chapter, for a DNA-like molecule (containing enough information to serve as a design template for even the most primitive forms of life) to have sprung by chance full-blown out of any conceivable collection of molecules is nonsense. The thing would have to evolve through literally millions of useless permutations before it could have any survival advantage at all. For more examples of evolution that seem to have a pre-determined end point, see chapter 18.

Another example of the complexity of life and its design is the immune system of vertebrates. It demonstrates the capability of a single cell, the one that supposedly assembled itself by chance in that magic swamp. The immune system is so complicated that scientists are still trying to understand it thoroughly. Even what is known is far too technical for purposes of this book. But a medical doctor who writes a popular newspaper column was recently asked to give a simple explanation of the immune system. He said he could give only an over-simplified version, which I'll paraphrase.

Cells have the ability to self-clean when their own cellular proteins get worn out. A special molecule marks the worn-out cellular material for recycling. Then other molecules cause the protein to unfold. Once it's straightened into the form of a filament, the protein is fed into barrel-shaped structures called proteasomes that chop them up into short fragments called peptides. The peptides are expelled from the barrels, and then they are either stored within the cell, or they are attacked and further degraded into amino acids for recycle.

For "enemy" proteins that come from outside the cell, the system is different. First the proteins are captured by sticky membranes surrounding the cell. The membranes then form dimples around the foreign proteins. These dimples become deeper and deeper until the proteins are surrounded by little bags that extend back into the cell. Then these bags are pinched off and disconnected, so they float in the cellular fluid, while the offending protein is completely sealed off from the cell itself. Now the bags and their prisoners are attacked by powerful enzymes that degrade them to amino acid for recycle.

Somewhere in the DNA molecule is a code that instructs all these molecules in exactly what part they are supposed to play, as well as instructions for the cell management systems that precisely control and choreograph this amazing dance.

What a superbly clever design! Yet until those dimples get deep enough to completely surround the offending protein and then pinch off the end, it's a useless and unworkable system. Why would it evolve through millions of years and thousands of steps, if there were no survival value in the steps, especially the early ones?

Now let's take a look at the bacterial flagellum. This amazing living machine was unknown to science until fairly recently. The flagellum is the mechanism many bacteria use to move around rapidly. There is a long shaft with a circular cross section that rests in a cylindrical bushing. The bushing is mounted in an opening in the outer membrane. And unlike anything else known in nature, that shaft rotates rapidly, both clockwise and counterclockwise, at speeds up to 100,000 RPM. The shaft connects through a rod, or drive shaft, to a little organ, which is the motor. No one knows yet just how this motor works, except that it is driven by a flow of protons, or hydrogen ions, rather than electrons. On the other end of the shaft is a bundle of whip-like flagella that serve as a variable-pitch propeller. Included in this mechanism are switches, sensors, controls, and memory. In other words, we have something very much like an outboard motor, except it is far more energy efficient. Furthermore, it is so miniaturized that eight million of those shafts could fit into the cross section of a human hair.

The question, of course, is: how could this machine develop by slow increments, or even fast increments? If it were missing any parts, it would be useless. A scientist named Michael LePage tried to finesse this question by saying the bacterial flagellum could have evolved because many of the proteins involved are in the cells performing other functions. This like saying the janitors in Oppenheimer's lab invented the atomic bomb because they were human and they were there.

Going back to the Minolta on Mars, what conclusion does the astronaut come to? Even Darwinians would agree that some intelligence made it, even though they didn't have any idea who or where that intelligence was. But what if that Minolta were alive as well as operational? The mainstream scientists would immediately deny that intelligence had anything to do with it. Logic flies out the window as soon as life is involved. And why? Because their God is Darwin, and Darwin put his stamp on life.

The bacterial flagellum is exactly the same as that living Minolta. We humans are pretty smart. We invented a lot of fine machines, including the outboard motor. But an outboard motor never invented or built itself. The flagellum is a living outboard motor. Is that any reason to reject out of hand the premise that it was designed? We don't know where that design originated, but neither did we know those things about the Minolta on Mars. Why shut our minds to the possibilities? Isn't an open-minded search for truth what science is all about?

Recently, Yale University hosted a conference of prominent scientists on a very surprising subject: Intelligent Design. This is something that could not have happened in the past 100 years, since Darwin became the god of naturalism. This and other signs of apostasy became so frightening to the naturalist establishment that they have been fighting back with every weapon (usually sneers and personal invective) at their disposal. But the evidence of miracles has become too strong, and the scientific world has begun to change. The truth may yet set them free.

16. The Brain

To exclaim on the wonders of the human brain is almost a cliché. Everyone knows and agrees that the brain is a marvelous organ. But do we really understand and appreciate just *how* marvelous it is? Maybe now, in this age of computers and supercomputers, it may be easier than in the past to take in the significance of this incredible thing we have inside our heads.

Essentially everyone who comments on the brain compares it to a computer. I have my doubts, but let's assume for the moment that the analogy is correct. Computer technology has come a long way in the last twenty years. Literally hundreds of thousands of mostly good minds have been working full-time on improving the performance of computers, and the results have been remarkable. Any saleable computer, including twenty-dollar hand calculators, can multiply two large numbers together faster than the supercomputers of the recent past. Your desktop computer (if any) can store, retrieve, and calculate more numbers in a second than you are likely to have any need for in the next hundred years. Children's games contain computers that make the famous original Cray supercomputer look like an abacus. Computers do a wonderful job of storing, manipulating, and returning typed words, like the ones on this page. They allow outstanding design, drafting, and drawing techniques. They open up the motion picture and game industries to extraordinary animation and special effects. They talk to us and translate for us. They control complex manufacturing processes. They allow us to fly rockets to other worlds, or smart bombs down smokestacks.

Because of the amazing numerical tricks that can be played with modern computers, many experts have been predicting that soon we will have "artificial intelligence", thinking machines that have all the prescience and sensitivity to nuance that a human has, with much more speed and capability -- something like the fictional Hal of the movie "2001, A Space Odyssey". We see frequent feature articles in respected magazines that celebrate some astounding computer achievement -- achievements such as a robot that can smile and frown at certain appropriate times, or one that can move about without crashing, or a chess-playing robot, even robots that can design other robots. We have computers (including the one this is written on) that have the ability to read in or "scan" a written document and then convert the characters to other written words that can be edited. We have computers that can understand spoken words, such as those at the other end of your telephone line that irritatingly tell you: "for English, press or say one, now". There is the remarkable "Smart" phone that can answer questions, identify music it hears, and see the person you're talking to. And there are animation computers that can make any scene the mind can conceive of into visual reality. These are cause for great jubilation as being signs of the coming humanization of computers.

But let's look a little more deeply into this business. Not one of the computer tricks people become so excited about is any more than a whole lot of electronic switches set (programmed) to do exactly what they are told to do by the human programmer. The computer that recognizes the spoken word "one" is merely analyzing the sequence of frequencies involved in the word, in strict compliance with the clever instructions of its hard-working programmer. The robot that gives a metallic smile when certain actions appear before its carefully designed sensors is likewise merely doing precisely what it was told to do by its program. The robot-designing robots are carrying out even cleverer instructions, including commands to use randomly-generated "ideas" in their work. Artificial intelligence is well named, especially the part about "artificial". There is nothing intelligent about it. When a machine "learns" it is simply responding to complex human commands to document and analyze the data it receives and to make certain kinds of carefully thought-out (by humans) adjustments in its previous instructions.

    Computers and their tricks are a wonderful testimony to the capacity of the human mind. No computer ever invented itself or assembled itself by chance. Any improvements a computer makes in future computers are at the express directions of the geniuses who invented and developed them. A simple example is the spell checker on your word processing program. When you first start working with the program, assuming you use words of more than one syllable, your spell checker will drive you crazy telling you it never heard of this word or that. But slowly, as you tell it to add the offending words to its dictionary, it seems to become steadily smarter. Eventually, it will know almost as many words as you do, not because it "learned" them but because you directed or programmed it to set its little switches so as to get off your back.

    No computer can possibly think an abstract thought, or feel an emotion. Except under the complete and exact direction of a human operator, it cannot paint a picture or compose a symphony or explore an atom. It cannot wake up one morning, spontaneously throw open the shutters, take a luxurious breath, and of its own volition write a beautiful poem. A computer is just a machine, like any other machine. No matter how complex the jobs it does, it can do only what it is told to do. If some of the things it does partly mimic what human minds can do, that doesn't mean it is thinking. Many machines lift heavy weights, but that doesn't imply the machines have been working out at the gym.

All right, now that we've suitably trashed computers, let's get back to the human brain. Within that small blob are abilities that make the most gigantic and sophisticated computers in the world look like Tinkertoys. For one thing, the human brain can invent and develop - - computers. Here's a very small, partial list of additional capabilities. The brain can:

a) Control the incredibly complex workings and interactions of the physical body, all automatically and without interrupting the conscious mind.

b) Take electrical and chemical signals from the optic nerve and turn them into panoramic, three-dimensional, color moving pictures that correspond exactly in size, shape, and position to the actual world.

c) Process the data from the sound waves emitted by the mouths of other humans and translate them into an entire language of thousands of words. Furthermore, it can instantly analyze a sentence full of mumbles, mispronunciations, run-together words, missing words, and cleared throats, and understand exactly what the speaker intended. (To be entirely honest, I have to admit that some of the students I teach are so mush-mouthed that even I can't understand them.)

d) From an almost complete blank at the beginning of its life, learn the above language simply by listening to it.

e) Be capable, without any outside direction at all, of thinking original thoughts, of looking beyond the images it sees to explore the causes and reasons for what is and what can be imagined.

f) Create, perform, and appreciate music, literature, art, and even philosophy.

g) Feel love, and all the exalting and bitter emotions associated with it.

h) Be aware of powers greater than itself, and attempt to understand them and reach out to them.

i) Dream.

j) Practice altruism and often true sacrifice.

k) Lose part of its mass through injury or disease, and yet somehow retrieve much or all of its lost memory or functions from elsewhere and transfer them to the remaining portion of its substance. In spite of massive efforts, no one has yet figured out how it does this.

This list is obvious, and goes on and on. The point of it is that no computer (and I maintain that no *conceivable* computer) can do any of these things, or even approach them. In fact, as one studies the abilities of the brain, especially the human brain, the more difficult it becomes to believe that the brain is comparable to a computer at all.

The brain contains millions of nodes called neurons that are connected to others through a massive network of nerve fibers. One of the features of a computer is a similar network of connections that make up the part called "volatile memory". Yet no matter how huge that network is made, the computer is still a machine, and can not do any of the tasks (or hundreds of others) described above. Obviously, something else is at work in the brain. What is that something? So far, very few scientists have faced this question head on. What many of them do admit, however, is that they are baffled by how the brain works. It seems impossible that such a small lump of living matter can do what it does, especially when it comes to emotions, abstract thinking, and the processes of creation and invention. Scientists know where in the brain many of these functions occur, but they have no idea how. Until someone can explain it and duplicate it (something that almost certainly will never happen), the brain must be considered a miracle.

17. Other Senses

We recognize five senses, sight, hearing, feel, taste, and smell, because we happen to possess them. Sometimes we are amazed at what appear to be other senses demonstrated by various animals. We ask: are they reading our minds, or each other's minds? How do they know what the weather is going to do, or that an earthquake is about to start, or a tsunami is coming? How does a dog lost hundreds of miles from home find its way back? How can some dogs locate cancer in a human or sense an imminent seizure? Always, it seems, we tend to explain these things on the basis of an exaggeration of the five senses we happen to have. Scientists are especially prone to such rationalization.

One of the most incredible of these phenomena, and the most difficult to explain, is migration. Every single eel in the Americas and in Europe is believed to be born in the Sargasso Sea, that huge, rotating, vegetation-clogged patch of Atlantic Ocean between Bermuda and the Bahamas. Their parents stay there to die, but the little elvers somehow find their way back to where their parents came from. The European eels stay with the Gulf Stream and head northeast, while the American eels turn off into various westward currents and head northwest. Much of their life cycles remain a mystery, but many scientists believe that they find the rivers, the streams, the very creeks their parents left before the elvers were born. They go through amazing hardships to reach those areas. Later, when they are ready to spawn, they will cross wet grass if necessary to reach the ocean and the Sargasso Sea. No one knows how they accomplish these feats, but it certainly isn't by use of the five senses we know about. Sometimes a scientist will try to invoke the sun, the stars, the currents, the temperatures; and maybe they're right. But let's see you crawl out of your bassinet floating in the middle of the ocean and try to figure out how to get to Granny's house. Especially without language of any known kind. It would be miraculous. Almost as miraculous as a tiny, orphaned elver swimming through the trackless ocean straight to his ancestral home.

The long annual migration of some birds is a special case. For example, flocks of snow geese will take off from the Arctic regions of northern Canada and fly over the western deserts to a specific spot in northern Mexico for the winter. Children are taught that the snow geese do this by various methods: a) following landmarks on the ground, b) noting the direction of the sun at noon, c) locating the North Star, d) performing dead reckoning, e) using an internal magnetic compass. All of these are methods we humans would use if we were making the journey, and all of them come straight out of someone's imagination. It's a wonder they don't teach that the geese listen to radio signals. One or more of the theories may well be correct, but the fact that there are so many of them shows that no one really knows.

However, one of the theories does seem to make some sense, and if true it raises an intriguing question. That theory is based on the fact that the brains of the geese contain a bit of magnetite, which is an easily magnetized iron oxide. The speculation is that this lodestone is used to guide the geese along the earth's magnetic field. If this were true, it would raise the question discussed in chapter 15: how could such a thing evolve? Of what possible evolutionary advantage could a piece of magnetite be in the brain of a migrating bird (or anything) until the brain was wired to detect and transmit the magnetic signal? On the other hand, what good would it do to have such wiring without the magnet to produce the signal? The magnet and the wiring would have to evolve simultaneously in the same mutant bird. It's not quite as far-fetched as the bacterial flagellum, but it's still absurdly unlikely. Shall we say miraculously unlikely?

There is a well-recognized phenomenon, discovered in the early 1980's, so strange and inexplicable that scientists are quick to admit they have no idea how it could happen. It is the spawning behavior of many coral reefs. About 4 to 8 days after the full moon in August, about two to three hours after sunset, at precisely the same moment, nearly all the coral polyps on a reef, no matter of what species, release their eggs and/or sperm into the water. The number of these little packets of new life runs into the trillions, and clouds the water that evening. Usually the little bundles contain both egg and sperm, which rise to the surface, fuse, and form microscopic larvae. Sometimes only sperm cells are released, some of which are returned to the parent coral to fertilize the eggs. Either way, the few larvae that avoid being eaten drift along for several days or even weeks, sometimes for great distances, until they find a promising spot such as a rock or a shipwreck to start a new reef colony. There they swim down, attach themselves, and begin asexual reproduction. The questions are: how could an entire reef, including the huge Great Barrier Reef, communicate throughout its length the exact moment when all the individual polyps would shed their eggs and sperm. And another question is: how could those tiny, eyeless larvae recognize a suitable substrate for their new reef?

No doubt there is an explanation for all this, and maybe someday scientists will figure out what it is. But no matter what the answer, it shows there are aspects of communication that do not use the normal five senses. Leaving -- what?

The coral phenomenon leads directly to the next topic, which I bring up with trepidation: extra-sensory perception (ESP). My hesitation is because many people on both sides of the dispute about natural versus supernatural do not believe ESP exists. So much nonsense has been spouted in the name of ESP by science fiction writers, as well as mediums and other assorted quacks, that people tend to equate it with men from Mars or "Chariots of the Gods". But in fact ESP exists. It has been shown to exist by careful scientific studies at Duke University and elsewhere. I know it exists because I once experienced it personally and dramatically (that's another story). The problem with ESP in humans is that it is extremely weak and unreliable. Most people cannot experience it at all, some have an occasional event, and some can do it much better than most; but no one can just walk around reading all the minds they see. When it does occur, it is most often at close range between compatible minds, and is quite limited in scope. Many identical twins seem to communicate wordlessly all the time, apparently without realizing how astonishing that is to ordinary mortals. And once in a while a message is transmitted somehow from one mind to another over a long distance.

How does such a communication system work? It certainly has no basis in the five ordinary senses. The electrical signals emitted by the brain induce a slight electromagnetic field, but it is far too small to extend more than a few feet, if that. There seems to be no known physical phenomenon that could explain ESP. This mystery, in fact, is the reason why so few scientists allow themselves to believe in its existence, in spite of hard data proving it. Some of them actually say that in spite of legitimate experimental evidence, ESP cannot exist because there is no explanation for it. There is an explanation, however, no matter how distasteful scientists find it. And you know what that is.

18. Evolution Toward a Goal

In previous chapters, we've gone into some of the weaknesses in the naturalistic, random evolutionary model of life and of human existence. Many scientists have come to acknowledge these weaknesses, and to admit that, in contrast to the puffed-up "certainty" of the recent past, we do not know how life started. Some have reached the point of saying publicly that life could not have come about without some kind of special creation. But having said that, a sizable percentage of even those enlightened souls still believe that once life started, everything then evolved by natural selection from that cell to -- us. They can develop very plausible scenarios for each step along the way -- that is, if you think a scenario with no data behind it is plausible. But unfortunately for those who rest their hopes on natural selection, there are a few glaring facts right in front of their faces that point up the illogicality of using natural selection alone to explain them.

Put into its simplest terms, natural selection works like this: Due to a mutation or other natural variance, a difference occurs between parent and offspring. Let's say, as a hypothetical and very extreme example, all moose have short legs. Then one day a freak moose is born and soon develops longer legs than its parents. If those moose live in an environment where the freak moose bumps his head continually on low-hanging branches, he will probably knock his brains out before mating and his long-legged gene will not be passed on. However, if the moose live in snow up to their bellies half the year, the long legs might be an advantage in getting around. The new-style moose finds a female (one who is not too fussy about mating with a freak), the offspring also inherit the mutated gene, and before you know it the race of long-legged moose have prospered and the short-legged version have wallowed themselves into oblivion. However, there is a lot wrong with this theory, especially when you use it to explain the evolution of an amoeba into a monkey.

The critical feature of natural selection is that for a mutation to be successful, it must confer some survival advantage to the species. Yet in many cases the evolution toward an advantageous goal must have gone through many stages having no advantage; having, in fact, a disadvantage. Features appear that have no function at all until some other features appear, and those other features likewise have no advantage in themselves. It is as if the evolutionary steps were aimed at a future goal that only some kind of intelligence could have been aware of. In addition to magnetite in goose brains, the bacterial flagellum, the human immune system, and the DNA molecule already described (Chapter 15), the following are some more creations that appear to be aimed at a future goal.

A. The Eye Lens

According to some Darwinists, the first life was a single cell (see chapter 14 for some comments on *that* nonsense). The single cell has no eye at all. So somehow the eye had to evolve from nothing through mutations and natural selection. But as you know, the eye is a very complex organ indeed. At the very least it must have a collection of light-sensitive receptors. Those receptors must be able to produce electrical signals of a strength determined by the intensity of the light they receive. To act as a true eye it must have a lens to focus onto those receptors the light that reaches the eye. It must have a multi-channel optic nerve that can carry all those signals from the array of receptors to the brain. It must have a region in the brain that can decode all those signals and convert them to a mental image of what the eye "sees". There's a lot more to the eye, but a primitive eye might be able to get along without them. However, the features I've listed are absolutely necessary for an eye to work at all.

Let me say at this point that there are eyeless animals that can detect light with light-sensitive organs on various areas of their bodies. But detecting light is a long way from having a functional eye.

So how could a functional eye evolve? Certainly we can't be expected to believe that some eyeless, primordial creature one day gave birth to a super-baby with a fully formed eye or two. Even the most rabid Darwinite won't go that far (at least, I don't think they will). Since evolution by natural selection is supposed to be a slow, gradual process of minute changes, each change lending some survival advantage to the species, we must visualize something very slight as a starting point. Let's start with one of the above light-sensitive organs. Over time, natural selection might improve the organ to the point where it can differentiate intensities or even color. Still, it's not an eye. Somehow, it would have to evolve a lens system to focus light onto a large array of light-sensitive organs. But now evolutionists come to a big problem. What would be the driving force to evolve a lens system? There might be survival advantages to slowly improving the light-sensitive organs, but what advantage would accrue from a primitive, rudimentary lens that didn't do anything? What would cause a lens even to begin its evolutionary journey? It would seem that the god of Random Chance would have to be able to think and plan ahead before any such adaptation could even begin to form. Although the existence of an intelligence behind the eye's design seems to fit the facts, such a design does not square with our scientific "laws". It is not explainable by science or by human reason at all. That, if you recall, is the main definition of a miracle.

A beautiful (or should I say ugly) example of the hopeless desire many scientists feel to "prove" their redeemer Darwin correct is a recent article in Scientific American by an Australian scientist named Travor Lamb. The article is called "Evolution of the Eye", with a subtitle of "Scientists now have a clear vision of how our notoriously complex eye came to be." After describing that same complexity in excruciating detail, he then proceeds to make an idiot of himself. First, he uses such words as "may have," "could have," "probably," "conjectured," "might," "I proposed," I conjectured," etc. literally dozens of times in a short article. At least he was honest about how murky that "clear vision" really was. Second, he actually resorts to the discredited doctrine of "recapitulation," in which an embryo supposedly goes through all the stages of its ancestor species in the course of development. The fact that it's been proven false does not seem to deter the Darwinian true believers. And finally, when it came to the meat of the subject, the formation of the lens, he describes the formation of the eye during embryonic development of a modern lamprey eel: "--- the lens forms, originating in a thickening of the embryo's outer surface ---." He then goes on to say: "It seems likely that a -- similar sequence of changes occurred during evolution."

Think of the massive illogic of this. He observes a lens (which he earlier had called an "invention") developing in a modern lamprey eye, under the explicit instructions of the creature's DNA, and then uses that observation to propose the same process caused a lens to invent itself, with no DNA instructions to help it along.

Scientists like this make my job easier.

B. Spiders and Their Webs.

During a period of twenty-five years in the mid-twentieth century, large groups of outstanding scientists working for DuPont finally developed a fiber ("Kevlar") that was stronger than steel of the same cross-section. It is used for bullet-proof vests, among other things. The process for making this material is almost unbelievably complex.

Spiders spin webs from fibers that are tougher than Kevlar, in some cases up to ten times as tough. The chemistry and engineering of this process in the spider is so complex and sophisticated that an entire book could be written on the subject. The raw material forms within their bodies and is pulled out of spinnerets, which are small-diameter extrusion orifices, the same as the spinnerets used by human fiber manufacturers. Some spiders have up to six spinnerets producing different kinds and diameters of web material for various functions. For example, sticky threads catch the prey, non-sticky threads allow the spider to move about the web without getting caught herself, and larger-diameter threads are used as primary construction members.

The process of making an orb web is a kind of miracle in itself. Here is a tiny, nearly blind animal that by feel scopes out the area to be covered, determines the attach points available, floats out an initial piece of silk to connect two points (using wind as the motive force), and then builds a complex structure that is strong, almost invisible, deadly, and incidentally beautiful. After a few hours, many webs become useless because the glue has degraded, or a large insect (or a careless hiker) has damaged it, or there's been a hard rain. So the spider just eats the web, recycles the protein, and spins another web.

Some spiders, of course, lack an esthetic sense and just throw their webs up in your garage any old way. Let us ignore these philistines.

In spite of its tendency to get tangled in your hair, an orb web is one of the most beautiful objects on earth, especially when sparkling with drops of dew.

Some spiders pull their web back, stretching it like an archery bow string, and leaving it cocked. When a mosquito approaches, senses the web, and turns around to escape, the spider releases the web, which springs forward and nabs the mosquito. Other spiders attach the lower part of their web to the surface of a flowing stream, using surface tension as a connecting force. Humans have not yet figured out how to do that. Yet another type creates a net, which it throws over its prey. One particularly amazing spider, the bolas, creates a single strong thread at the end of which is a large drop of sticky fluid. The spider dangles the thread with the drop until a moth comes along. The reason the moth comes by is that the spider has emitted an attractive pheromone that makes the moth think the strange creature with a bola hanging below it is actually a lovely female moth disguised as a strange creature with a bola. So when the lovelorn moth comes near, the spider winds up and throws the bola, snaring the moth. So much for romance. There may be a metaphor here.

Anyone who has been intimately involved in the manufacture of ultra-strong, flexible fibers has to wonder how such a material could develop inside a spider without some kind of directing principle. What good is a strong fiber material without a spinneret to form it? And what good is a spinneret without something to spin? How could a super-strong material evolve without going through many steps of useless weak material? What would cause something like web spinning to start evolving in the first place? It is as if the evolution were aimed at a distant goal, defined and directed by some prescient intelligence.

C. <u>Bombardier Beetle</u>

This incredible insect has long been on the front line of the argument between advocates of slow evolution and those of intelligent design. That's because the little ½" long beetle shoots *boiling water and steam* at its enemies, water that contains a toxin just to make it interesting. It can direct this blistering stream backwards, forwards, or to either side. The way it does all this should be of particular interest to anyone who knows anything about chemistry or chemical engineering.

Its body contains a chamber in which it stores a mixture of concentrated hydrogen peroxide and hydroquinone. It makes these reactants ahead of time in separate facilities and mixes them in the storage tank. When threatened, the bug opens a valve on the storage tank and feeds the mixture into a thick-walled reaction chamber that contains water and catalysts. The catalysts cause the hydrogen peroxide to break down, releasing oxygen and a lot of heat. The oxygen, catalysts, and heat cause the oxidation of the hydroquinone to form toxic p-quinones. The temperature rises to the boiling point of water, and the resultant pressure increase closes the check valve into the mixing chamber. Now the weapon is ready, only a second or two after the threat. The beetle swivels its abdomen around and shoots the doomed attacker with a noxious, boiling stream, 20% of which is steam. As an added feature, in some beetles the exit valves open and close at a frequency of 500 cycles per second, which apparently keeps the beetle from being jetted away by its own recoil. Another interesting feature is that when the beetle fires forward by swiveling its abdomen under its body, the steamy mixture goes right between and through its front legs; yet the beetle is not injured. The same cannot be said of its attacker.

Evolutionists need to answer some rather basic questions about how this creature came to be. In addition to the myriad obvious questions, such as the presence of hydrogen peroxide in one container and the reaction catalyst in another, there is one overwhelming question that they cannot answer. In order to fire the potent liquid out, the water must boil. Otherwise, there would not be enough pressure in the reaction chamber. Slow evolution of the catalyzed hydrogen peroxide decomposition would have no survival advantage; a dribble of warm water on the beetle's rear feet would just make the spider laugh before wrapping up his prize. But before creating that boiling mixture, the beetle would first have to evolve the thick-walled combustion chamber, and the check valves that keep the beetle from burning herself up, and the controlled high-pressure nozzle with its intermittent opening cycles. A chemical engineer could plan and do these things before the reaction was first tested. But an engineer is not random evolution. At least most of them are not.

This little insect continues to cause great grief to Darwinian evolutionists. Many scientific papers on the subject will include a defensive attempt to prove the bombardier beetle could have evolved by natural selection. Essentially all of these efforts have serious weaknesses. One gentleman named Isaac has dreamed up a set of fifteen evolutionary steps which he says might produce a bombardier beetle. Most of the steps address hydroquinone, which has nothing to do with boiling water. Working almost entirely without the benefit of data, he spends a lot of time trying desperately to show that the non-Darwinian purists could be wrong. His fifteen hypothetical evolutionary steps (several of which involve large, unexplained leaps) prove nothing whatever, except that his elaborate scheme seems to be directed toward a preconceived goal, something he definitely did not want to say.

An even more egregious example is a young English gentleman named Dawkins who seems to be making a living going around trying to convince English boys that God is named Darwin. His "proof" that evolution produced the bombardier beetle is that more dilute solutions of hydrogen peroxide produce only a warm liquid. How this does anything except prove him wrong is not entirely clear.

The bottom line is that all the efforts to discount the miracle of the Bombardier ignore the pressure question above. To reiterate: the beetle's attack mechanism cannot work unless the liquid boils, and it cannot boil until a thick-walled reaction chamber is formed and the correct valving arrangement is present. Evolution that looks ahead does not fit the laws of science.

19. Biochemistry

One of the leading originators of the Intelligent Design movement is a prominent biochemist, Dr. Michael J. Behe, a professor at Lehigh University. He wrote an outstanding book called "Darwin's Black Box" in which he points out that in the middle of the 20th century scientists held a series of interdisciplinary meetings aimed at developing an integrated position on Darwinian evolution. The result was called Neo-Darwinism. Unfortunately, the meetings did not include the science of biochemistry, because biochemistry did not exist at that time. If it had, and if the facts Dr. Behe discusses in his book had been known, the outcome of the meetings might have been considerably different. That's because when scientists finally understood the mind-bending complexity of the biological processes that occur on a molecular level, the idea of random evolution became absurd. If you read his book, you will find that everything you have seen about life so far in the book you are now reading has been over-simplified by orders of magnitude.

Because he is an active member of the scientific community, as well as the even more intolerant academic world, Dr. Behe has to tread lightly when using the word miracle. However, one could ask: within what scientific law does an invisible intelligent designer of living organisms fit? It does, however, fit the definition of miracle.

"Darwin's Black Box" discusses several organisms that exhibit what Behe calls "irreducible complexity," which includes "evolution toward a goal" (chapter 18) but takes in much more ground. In fact, nearly everything in the biological world, when considered from a molecular viewpoint, is so unimaginably complex that almost no one has ever made a serious attempt to describe what evolutionary steps could possibly explain their existence. Darwinists merely state that an organism or a biological system evolved, without saying how.

The complexity arises, as Behe says, because the organism in question requires the coordinated operation of several simultaneously "evolved" complex features in order to work at all. You have already seen some of these from an engineering standpoint, such as the bombardier beetle, the bacterial flagellum, the immune system, and the eye. This chapter will present several more, based on Dr. Behe's book. These systems are so complicated that he simplified them for his readers. But even the simplifications are so complex that I will further simplify and paraphrase.

A. Antibodies.

Chapter 15 included a rough description of the human immune system, but it did not include the crucial role of antibodies. Most of us non-biologists think of antibodies as the agents that dash around mopping up invasive entities in the body. But this is incorrect. Antibodies are like detectives that identify the foreigners, capture them, and turn them over to the executioner. The miscreants don't get a trial. But here's where the story gets interesting.

To capture an invader, an antibody must fit exactly into a corresponding shape on the invader. Since there are at least one hundred thousand possible shapes on foreign entities, there must be at least that many differently-shaped antibodies ready to spring into action. And, in fact, in order to increase the chances of finding an invader quickly there are billions of different antibodies in the body. This diversity is the result of an incredibly clever randomization process involving mixing and matching genes from four different gene pools. A PhD statistician could do no better.

An antibody is a group of four amino acid chains formed in the shape of a "Y". At the ends of the two forks are random shapes, which are the binding sites. The tail end of the antibody is hydrophobic (water-hating) and stays down inside a special "B" cell, which is a factory for antibodies. The "B" cell roams around with the "Y" of its antibody sticking out. Each of the billions of factory "B" cells has its own unique antibody. By the time an antibody and its attached factory find and capture an invader, the enemy has multiplied many times, and the lonely antibody has little chance of defeating the army. So it calls for reinforcements.

To do this, the "B" cell, working in complete harmony with its antibody, swallows the enemy and chops it into pieces. About that time, another cell called a messenger-T cell comes along and attaches to the B cell. The B cell presents a piece of the enemy to the T cell "for consideration" (as Behe says). If the T cell confirms the diagnosis, it sends a message back to the "B" factory cell to grow and reproduce. Many other "B" cells form and begin pumping out large quantities of antibodies that are identical to the original one that found the interloper, except that the tail ends are no longer hydrophobic. Thus the newly cloned antibodies can leave their ponderous "B" cell friends and roam much more freely. However, they stay close to their own factory cell.

Another event that increases the system's effectiveness is that the DNA in the factory cell begins to mutate to form antibodies that are able to bond even more tightly to the foreign molecules. Soon there is a huge swarm of antibodies with just the right configuration to bond with the particular pathogen that invaded the body.

Eventually none can escape being captured. But that's all the antibody can do. The actual execution is carried out by an entirely different mechanism.

At this point, Behe's "simplified" description becomes so complicated that I won't even try to paraphrase it, thus sparing both of us brain fever. Suffice to say that a large group of proteins attaches itself to an antibody that has captured a foreign cell. It ignores antibodies that are still moving about free and also antibodies that are attached to "B" cells. This smart move avoids wasting proteins or destroying the wrong cells. Then a long, complex cascade of reactions occurs, which culminate in the construction of a tube which punches a hole in the invading cell. Water pours in and bursts the doomed cell.

And we are supposed to believe that all this evolved by random chance? Please.

Biochemists now understand nearly all the mechanisms by which the immune system works. But one question is so far unanswered. Antibodies are exposed to a never-ending swarm of cells that belong to its own body. How do they know to ignore them?

One final comment. Various hysterics in our society believe that one molecule of a poisonous substance entering our bodies might end up killing us. As chemical analyses get more sensitive, tiny amounts of more and more evil chemicals are detected, leading to ever more stringent standards. These people seem to forget about the immune system. Small amounts of anything are no problem for this amazing army to defeat. Only when massive amounts of a pathogen or a poison enter the body is the immune system overwhelmed, at least temporarily. But eventually it almost always catches up and wins the battle. Sometimes, if it detects a heat-sensitive bacteria or virus, it raises the body temperature to kill the critter. No matter what the invader is, there are already in place antibodies of just the right shape to capture and kill it, even if they've never seen it before. The hysterics need to take a deep breath and calm down.

B. Cilia

Cilia are the little hairs that act as paddles to help cells to swim. They also line the respiratory tract to help move mucus out from where it isn't wanted. Until biochemists came along with their electron microscopes, no one knew how they worked. Most of us, if we thought about them at all, assumed they were controlled by the organism to which they were attached, somewhat like legs on a millipede. Not so. All the organism does is provide raw energy.

When cilia are removed cleanly, and an energy molecule called ATP is provided, the cilia paddle away just as if they were attached to their parent cell. Obviously they are self-powered. Careful examination of the cilia cross-sections with the powerful electron microscope shows there are small projections inside the cilia made of a protein called dynein. These are the motors that move the cilia; if they are removed, the cilia are paralyzed. Ignoring for a moment the seemingly miraculous ability of a protein to act as a motor, the real eye-opener is how the cilia produce the swimming motion.

A cilia consists of nine tubes stacked vertically in a circle. Each of the tubes is connected to the ones next to it by slack cords. Now a little dynein motor moves down inside the stack, grips two adjacent tubes, and moves one of them up relative to the other. This causes the slack cords to tighten. Continued motion by the dynein can only lead to one thing: the entire stack bends. Then the dynein pulls the adjacent tubes in the opposite direction, the cords tighten, and the stack bends the opposite way. And voila! We have a fishtail-like swimming motion.

Now, Darwin fans, please explain how random evolution could produce that combination of slack cords and clever motors that push and pull adjacent tubes. Of what use would any of these structures be if one of them were not there? Or if the mechanism were not fully operational?

C. Blood Clotting

You cut yourself. Almost immediately, blood at the site begins clotting to shut off the bleeding. A scab forms, and healing begins. Simple, right? You should know better by now. Let me just say that if chemistry and chemical engineering were half as complicated as biochemistry, I might have become a street sweeper.

The first thing that needs to be done after a cut is for the system to determine that a cut has occurred, where it is, and how big it is. Then it has to form a clot at that spot, and at no other. Finally, it has to stop forming clots, lest it clot all the blood and kill the patient. And it has to do it all automatically.

Let's start with the second main step, the clotting itself. The blood plasma contains a modest amount of a placid protein called fibrinogen. It has nothing to do but hang around waiting for a job. But when a cut occurs, another protein called thrombin attacks the fibrinogen and cuts off some small pieces. It turns out those small pieces were covering up some reactive spots on the fibrinogen, and with the coating gone the fibrinogen changes its name to fibrin and becomes a ball of fire. The only thing missing in this metamorphosis is a telephone booth. Now a whole swarm of these newly-activated fibrins rush to the site of the cut and begin connecting to each other at the active spots. The result quickly becomes a mesh, something like a spider web or a fish net, covering the area of the cut. The holes in the net are smaller than the blood cells, and soon the trapped blood cells form a clot and the patient is saved. It seems to me this whole step is about as miraculous as things get.

But then one must ask: why doesn't that aggressive thrombin attack fibrinogen in the absence of a cut, cause a big net to form everywhere, and plug up the circulatory system?

The clotting system all starts with a gore-loving protein called Hageman factor that rushes to the scene of the accident and sticks to the surface of cells near the cut. The Hageman factor then apparently sends out a signal somehow for someone to come look at this mess I'm attached to. Another protein called HMK obliges by cutting pieces off the Hageman factor. Just as was the case with thrombin cutting and activating fibrinogen, the Hageman factor is activated and rushes off to slice up someone else. This starts a long and complicated cascade of various proteins getting cut and activated, in some cases joining with other proteins to gang up on a third. Finally, the circle of cutting and activation comes around to thrombin, which until then was also just floating lazily without a job. The thrombin is cut and activated, immediately goes after fibrinogen, and the rest is history.

To stop the clotting before the whole blood system turns solid, two proteins attack and destroy some of the intermediate proteins in the foregoing cascade. Behe doesn't say so, but perhaps these attacks occur more slowly than the clotting reactions, so the latter don't get shut down before their jobs are done.

Not only is this amazing dance far outside anyone's ability even to propose an evolutionary route to the business, every step of it is its own miracle. How does an inanimate protein molecule know to go to just the right kind of different protein and slice off the coating that covers its active sites? How do the active sites on fibrin happen to be in just the right spots to allow bonding with other fibrins to form a web, the holes in which are just the right size to stop blood cells from passing through? How could all these steps evolve simultaneously?

D. DNA Transcription.

If nothing else in this chapter convinces you that biological processes are crammed full of miraculous actions, the one I am going to describe now will surely do the trick.

First, you have to visualize DNA, the extremely long, double-helix molecule that looks like a twisted ladder and which contains almost all the information to make up a living creature. All along the length of the DNA are sections that constitute genes. Transcription occurs when a cell wishes to retrieve a gene for use elsewhere. (Even that previous sentence gives me a spooky feeling; how does that cell know it wants a gene? But let us proceed.) To get the gene, the cell must make an RNA copy of that portion of the DNA. (RNA is similar to DNA except it is only a single helix and it uses some different nucleic acids.) The cell calls on a long chain molecule called RNA polymerase. This material rides along the DNA like, as Behe puts it, "cars on a roller coaster," until it finds a marker that tells it the needed gene is just ahead. At that point, the polymerase stops riding the rails, binds tightly to the DNA, and starts doing a very complex and multifaceted job.

Consider an ant colony. When something invades an ant nest, every ant immediately starts to carry out its job: some carry eggs to safety, some move the queen, some repair the nest, and some go to war. Each has one task. But they have brains. After taking that roller coaster ride and finding the needed gene, RNA polymerase does a job so complex that many humans would find it difficult. First it opens up the DNA, separating the twisted strands for about 10 units. Then it gets inside the open part of the DNA and moves down, creating an RNA chain by matching to the DNA bases. After matching a pair and binding the appropriate group to it, the polymerase moves to the next position and matches again. But as it goes further down inside the DNA spiral, it causes the DNA ahead of it to begin getting too tightly wound. This is where the mind-blowing part comes in.

Another protein comes to the rescue. This one untangles the DNA. It does this by "cutting one strand of the tangled DNA, passing the uncut DNA strand through the cut strand, and resealing the cut." What!!??

Anyone who has untangled a garden hose will have to admit: here we have a mere protein that is smarter than we are. Or that is miraculously directed.

I could continue with paraphrases of other biological processes in Dr. Behe's excellent book, such as transport of proteins from one part of a cell to another or a more complete description of the eye, but I think you get the idea, and my headache medicine is running out.

One final comment: the movement that came out of Behe's book is called Intelligent Design. That's fine as far as it goes. But I don't think it goes far enough. As an engineer, I might be able to design and build a mechanism that could paint a car automatically. But then I could point to each step in the process and explain how it worked, what electronic signal moved what valve, what lever moved what brace, etc. The computer that controlled the mechanism would be an extension of my brain, so I indirectly controlled it all. It would be an Intelligently Designed machine under intelligent control. But what are the ways to describe how biological systems work, no matter how intelligently they are designed? How does a brainless, essentially lifeless protein leap into action and carry out its functions at the appropriate time and place? What makes a dynein grab a tube and pull it down, then push it up? By what means does protein A rush over to protein B and cut off its coating, activating it? These big molecules dash around acting as if they had brains.

If the design of my painting machine were analogous to these biological systems, I would simply say: "When the car moves to this point, a paint sprayer will come over and paint the fender, then an infrared lamp will come down and dry the paint," etc. This is a design, but it fails to specify how these things are to be done. It would not function.

I suggest that the term Intelligent Design is inadequate. It should be called Intelligent Design and Control.

Of all the miracles discussed in this book, I believe this may be one of the most significant. Even those scientists who continue to argue that biological design evolved in a Darwinian manner must surely admit they cannot explain how the resultant brainless design actually works.

20. Quantum Theory

This will be a short chapter because I don't understand quantum mechanics. The good news is that I'm in excellent company; many of the scientists who are experts in the field say they don't understand it either. Given some of the ideas that have come out of it, I'm not sure anyone does. In fact, a very famous modern physicist, Richard Feynman, has said, "I think I can safely say that nobody understands quantum mechanics." And he was one of the world's leading experts in it. The reason this chapter is near the back of the book is because some of the implications of the theory are so strange that if I discussed them earlier you might think I was crazy and you would not read any more.

The definition of the word "quantum" is "The smallest amount of a physical quantity that can exist independently, especially a discrete quantity of electromagnetic radiation." As mentioned in chapter 5, one of the main precepts of quantum mechanics is that light (and other electromagnetic energy) is both a wave and a particle. In fact, according to quantum theory (confirmed by experimentation,) matter, too, is both a wave and a particle. This fact adds another layer to the statement in chapter 3 that mass and energy are essentially the same thing.

Another apparent fact to come out of quantum theory is that nothing is what or where it seems to be. One of the two fundamental principles of quantum theory is that everything, at least in the realm of the very small, is based on probability. For example, electrons do not circle the nucleus of an atom in nice defined orbits like the planets. They do circle the nucleus in a series of nested "shells" that determine the atom's reactivity. But within each shell they are said to exist as a sort of swarm or cloud, and it's impossible to state exactly where any individual electron is. Instead of being able to say an electron is *here*, you can only say there is a 1% probability that at any one moment the electron is here, and another 1% probability that it's over *there*, and even a finite probability that it's up next to your ear, etc.

This uncertainty as to where an electron is located is one of the reasons various scientists give for why electrons, which are negatively charged, do not fall into the nucleus, which is positive and according to the well-known principle that opposites attract would be expected to attract the electrons. I will not attempt to explain this effect because I don't want you to share my headache. Suffice to say that as you add up all the probable positions of the electrons around the nucleus, the most probable position is somewhere other than the nucleus.

Another reason, much easier to visualize, is that the electrons are whizzing around the nucleus at nearly light speed, and they are kept there by centrifugal force. What causes them to whizz is, like so much else in the universe, a mystery.

There is a law describing the positional uncertainty called, not surprisingly, the Heisenberg Uncertainty Principle. This says (briefly) you cannot simultaneously measure both the speed and the location of an electron. The common way to explain this is that in order to observe an electron you must hit it with a photon. But that collision knocks the electron out of its path and/or its velocity. This is true but is not the real answer. There are complex mathematics behind it, but the simplest way to look at it is that the electron is not just a particle, it is also a wave. And depending on the moment in time that you look at a wave, it can be in any number of places.

In fact, it's not just electrons that have this dual nature, it's everything. All particles, protons, neutrons, photons, atoms, molecules, everything is both a particle and a wave. This is the other of the two fundamental principles of quantum theory. It can be experimentally proved, as I'll show in the next paragraph. But how can that be? How can it be visualized? Einstein and all the other great figures in physics were distraught when they realized the duality. They all have attempted to explain the effect, and all have fallen short. Some of the attempts have been reduced to metaphysics, which gets us very close to miracles.

Here is how one can prove that light is a wave and a particle, and it puts us into a new world of strangeness. If you shine a light through two side-by-side parallel slits and project the results together on a wall or a screen, you do not get two parallel lines of light. You get what are called interference patterns, which are wavy light and dark bands. Sometimes you see them on your television screen. They are caused by the light from one slit sometimes adding to the light from the other, and sometimes interfering or canceling the other's light, even though both stem from the same light source. Only waves could produce this effect. But then if you shine the light though only one slit, you would expect the interference patterns to disappear. But they don't. The patterns are still there. Scientists do not have a clue how this is possible, except to say that the light somehow exists in two places at once and is interfering with itself. Are you still there? Do you see why a chapter on quantum mechanics is included in a book about miracles?

Another, even stranger, phenomenon involving slits is that if you send electrons from an electron gun one at a time though a narrow slit and impinge the light on a type of screen that retains the light's image, you see that the light is causing minute spots on the screen, showing the light is hitting the target as a particle. But as you continue to shoot electrons, you find that the spots on the screen are forming wave patterns. Obviously, each individual electron is not travelling in a straight line, but is moving as a wave, while still retaining its character as a particle.

Quantum mechanics repeatedly operates on the principle that an individual particle or wave of matter or energy can exist in more than one place at a time. Often it can be in many places, up to and including an infinite number of places, depending on who's doing the calculating or the speculating. If matter can actually exist in multiple places simultaneously, how can it be explained on a physical or logical basis?

Speaking of miracles, the quantum nature of physical reality, such as it is, leads also to a phenomenon called "entanglement". Quantum theory predicted that when two particles, say two electrons or two photons, become entangled, they will retain their connection when separated. If I tried to explain how they got entangled in the first place, that headache would become a migraine. Anyhow, some scientists say they will remain connected even if they end up on opposite sides of the galaxy. Whatever move one particle makes will be reflected in a move by the other. Not surprisingly, even Einstein, that master of the bizarre, refused to believe it. As a consequence, he decided that quantum physics was invalid, and he spent the last twenty years of his career trying vainly to prove it.

But apparently he was wrong. In the early 1980's scientists using a laser shining through a crystal succeeded in producing an entangled pair of photons and sending them in opposite directions. Sure enough, they observed one photon reacting to the movement of the other. Recently other researchers have managed the feat with electrons and with actual small molecules. Just as was the case with the photons of light that exist in two places at once after emerging from a slit, no one has any idea how such a thing is possible. Some quantum experts have said it is "counter-intuitive", which in English means they haven't a clue. Either the universe is *totally* different from our perception of it, or we are witnessing a miracle. Or maybe both.

So much for quantum phenomena that are true and actually observable. Now, to complete the picture, we must look at what all this weirdness has produced in the minds of some people who have studied it too long. In the 1920's, some scientists began to postulate that quantum theory and probability principles dictated that if two outcomes of an action were possible, then both outcomes actually existed simultaneously until an observer could show otherwise. However, in those days scientists in general still had some common sense, For example, Erwin Schrödinger, the man who wrote the equations describing the wave nature of matter, proposed a thought experiment in which a cat is confined in a box with equipment that might or might not kill him. According to quantum theory, said Schrödinger, until the box is opened the cat is both alive and dead. Only when he was observed did the paradox disappear. Einstein also proposed a very similar thought experiment. Both men scoffed at the idea that a cat could be simultaneously alive and dead, declaring it nonsense, and used it to try discrediting quantum physics in general.

Unfortunately perhaps, quantum physics turned out to be true, at least in the world of the very small. It went from a being a mathematical curiosity to an actual, if seemingly miraculous, vision of microscopic reality. And at that point, because it was so difficult for even top-flight scientists to understand, science began to go astray. The dead-and-alive cat became reality to many physicists.

In 1956, a Princeton graduate student and science fiction buff named Hugh Everett wrote a thesis in which he claimed that for every action, there is an alternative action, and that the results of both actions actually exist. In fact, the alternative action would create a complete separate universe. He thought this would give him immortality, because at the moment of his dying, there is an alternative world in which he does not die. Obviously, since there is nearly an infinite number of actions going on by billions of people (not to mention animals, vegetables, and minerals), if a new universe is created every time there is a possible alternative action, the number of universes soon approaches infinity. Where all the matter and energy for these new universes is supposed to come from is not clear.

Now this crackpot proposal would not be worth mentioning, since who knows what a graduate student might be smoking? But the frightening part of the story is that Everett has become a hero to a significant contingent of quantum scholars. They actually believe it. Many of them now make didactic pronouncements that there is no longer any question about the truth of the "many worlds" idea. And these are the same people who say they cannot believe in miracles.

Once the door to the asylum had been opened, there was no stopping the inmates. After the militant atheist Hugh Everett died prematurely of drinking, smoking, and obesity, his daughter committed suicide so she could join him in a parallel universe. If the man had not disproved his own theory by dying, his teaching a child such a thing could be construed as child abuse.

Another belief that has become popular is that if one could enter a black hole, he could "tunnel" into one of those parallel universes (see chapter 9). The physicists who believe in these tunnels are not deterred by the fact that the gravity in a black hole is so great a human body approaching one would be torn to pieces by the difference in gravitational force between his head and his feet.

The fact that probability governed the world of the very small led many scientists to believe the proposal that if a rubber ball were thrown against a stone wall, there was a finite chance it would go through the wall instead of bouncing off of it. These enthusiasts are forgetting that rubber is a long-chain polymer, and the chances of even one molecule of rubber sliding between the tight-packed molecules of stone are not extremely low -- they are zero.

Another indication of the disconnect between the quantum world and the classical (or normal) world is that quantum theory requires there be an equal amount of matter and anti-matter in the universe. For that not to be true would be a violation of the laws of quantum physics. But, as reported in chapter 3, the anti-matter is missing. Thus we either have proof that quantum physics is wrong, or we have a miracle.

Some very recent physicists are now theorizing that space and time themselves may not exist. Even the existence of gravity is being questioned. After reading chapters 1, 2, and 9, you may tend to agree. In any case, let's hope the King does not wake up.

There is one mitigating factor in all this quantum craziness. As you have seen in many of the early chapters, the universe is so beyond human comprehension that it seems almost anything is possible. A miracle is a miracle. Maybe quantum mechanics is the ultimate miracle.

21. Altruism and Compassion

Dr. David Demick has pointed out some very interesting facts about the relationship between the plant world and the animal world. According to the Darwinian doctrine of survival of the fittest, the plant world should dominate the animal world. Plants have a much more complex and sophisticated biochemical mechanism than do animals. In the dark, they have the same basic life processes as animals; they take in oxygen and organic fuel (such as sugars) and convert them to carbon dioxide, water, and energy. But green plants also contain the remarkable chlorophyll system that when exposed to sunlight acts as a photoelectric cell to extract the energy of the light and reverse the oxidation reaction. In the sunlight, they take in carbon dioxide and water, and produce oxygen and sugars or other carbohydrates. The animal world is entirely dependent on plants to regenerate the oxygen and carbohydrates we use up. If plants operated only on the photosynthesis process, as we were taught in school, we could say that plants and animals were in perfect synergy, each depending on the other to recycle its waste products and regenerate its fuel. But in fact plants are potentially independent; it may well be that they have no need of us at all. Presently, they do not consume as much oxygen as they produce, but that may be an equilibrium phenomenon, caused by the large amounts of carbon dioxide produced from other sources. If there were no animals, no volcanoes, no forest fires, no rotting vegetation, no other sources of carbon dioxide, it seems likely that plants would adapt by oxidizing more carbohydrates to form their own CO_2 for photosynthesis.

So with a vastly superior biochemical system, why is it that plants do not dominate us? One might say that their processes move too slowly. But in fact, internal plant processes are extremely fast. It is simply that they are designed to stay generally in one spot, their root systems extracting nutrients from the soil. It's just lucky for us that plants do not have a brain and a means of locomotion; if they were able to dash about at will, they might be hunting and eating us instead of the reverse.

But is it just luck? The animal world contains just about every permutation and combination of design possible. In a world that contains the three-toed sloth and the duck-billed platypus, it would be perfectly conceivable to have an animal with a dual system of respiration and photosynthesis. Darwinian evolution almost edicts that such a superior being should have developed. Yet, except for some one-celled organisms, there is no such thing. Plants sit quietly, docilely allowing us to do as we wish with them, including kill and eat them. Their dual life processes are used exclusively to produce both carbohydrates and oxygen, as if they were designed specifically to make life possible for us and our animal associates. If plants did not so carefully fine tune the atmosphere, all the other wonders described in previous chapters would be useless; we would not be here. Furthermore, plants produce most of the vitamins, proteins, minerals, and nutrients we need to stay healthy.

There is more. One might rationalize plant biochemistry as simply an adaptation that benefits the plants, with coincidental, unplanned benefits to us. But then we come to the incredible finding that, taken as a whole, plants contain chemicals that are curatives for most of the ailments known to man or beast. Many of these chemicals impart no known survival advantage to the plant. They are simply there, as if placed there for our use. How and why did such an apparently altruistic branch of life develop? To Darwin, this question was an "abominable mystery", and it remains so for evolutionists to this day. It *is* a mystery, but it's not abominable. It's a miracle.

And that brings us to our final subject: altruism (and its companion compassion) in humans and some other animals. The anthropologists tell us that these noble emotions are not really noble; they are merely an adaptation having great survival value for the family, the tribe, or the herd. The anthropologists are, in fact, very likely right (although it is a little difficult to divine the survival value of a widow who mourns and weeps for two years after the death of a beloved husband, or even a herd of elephants who delay for days their crucial search for food and water to mourn the death of a herd member). But even accepting the obvious advantage to the family of unselfish and compassionate behavior among its members, how does one explain such behavior when applied to total strangers? Let's look at an example.

You are an American businessman visiting Hong Kong to close a deal. You are walking along the street when a poor Chinese woman is hit by a car next to you. She is lying in the street writhing in great pain. What do you feel, and what do you do? Well, unless you are a compete swine (and there are a few of those), you feel compassion for her. You rush out and help stop traffic so she won't get run over again. If some one else has done that, you may well kneel down and try to comfort her, shield her from the sun, stop the bleeding. Only when you're sure others have the situation in hand will you leave her, and even then you may feel you want to stay and help more. You may be missing the beginning of a meeting, but at this point a meeting is secondary. Yet this is a person of another country, another race, another economic class, a woman from another world entirely. She is nothing to you.

Now, just how does your sympathetic and unselfish behavior toward her impart any survival advantage to you? If you were a creature of Darwinism, you would not even slow down; you might turn your head in curiosity, feel nothing but disgust for the noisy, bloody sight, and walk ahead to your meeting. That would be the optimum course of action for you *and* your family. So why did you ignore that imperative? Why did you pity that foreign woman, and try to help her?

The great C. S. Lewis wrote an entire book about this and related subjects. I will summarize it by saying that the existence of compassion and unselfish behavior toward strangers, even foreigners, is proof that something beautifully illogical has been given to us. There is something fine and noble that has been implanted into our minds and our spirits that makes a mockery of the concept that everything is based on survival of the fittest. We did not give this gift to ourselves. Darwin's evolution did not give it to us, and neither did Freud's Oedipus Complex. It came from somewhere else, from something that science and reason cannot touch. Its presence in our lives is proof that, among all the physical miracles around us, we too are a miracle.

22. The Convergence

The supposed conflict between science and religion sometimes seems a hopeless holy war between two fanatical sides who refuse even to consider the other's case. If anything, the scientist's "religion" seems to be gaining in power and influence. From the days of the Scopes trial when naturalists were asking only for "fairness" in allowing evolution theory to be taught in schools along with creation theory, we have progressed, or rather descended, to the current situation where the evolutionists refuse to allow anything but Darwinism even to be mentioned in schools. The idea that miracles have ever occurred, much less that they are all around us in plain sight, is absolute anathema to most scientists and other secularists. If a reddish creature with horns and a tail were to appear to a typical scientist and turn his wife into a chicken before his eyes, he would refuse to believe it was a supernatural event. So how is he expected to believe that something as ordinary as a spider is a supernatural creature? Almost everything religious people believe seems to run counter to what scientists believe on the same subject.

Yet there is a fact that promises to bring these two warring camps together, much as scientists may dislike it. It is a belief that is absolutely basic to the doctrines of both religionists and scientists, and it is exactly the same for both. That belief has to do with time, space, and the Big Bang. Devout, knowledgeable Christians, Jews, and Muslims all believe that everything, not just matter and energy but time and space themselves, were created together, all at the same time. This belief is fundamental to honest faith in a creating Deity.

I have already shown in chapters 1 and 2 that from a scientific, naturalistic viewpoint, based on known "laws" of physics, the idea of time or space starting or ending at some point is absurd. So is the idea of infinity; the notion that they *didn't* start somewhere. The paradox these ideas lead to was the basis for my contention that time and space themselves are proof of the miraculous. But the fact is that most scientists believe exactly what the devout religionists believe. They have developed supreme faith, based on their data, that matter, energy, *time*, and *space* were created together in the instant of the Big Bang. If scientists truly accept the concept that both time and space started at a single point and a single moment, then they are not only proclaiming a miracle, they are proclaiming exactly the same miracle as religions proclaim. The war is over. The truth has prevailed. Most of the defeated troops may fight on from their caves and bunkers, unwilling to accept surrender, but eventually every intelligent scientist must face the fact that he believes in something that is "scientifically" impossible, something that is miraculous. And when he does, he will find his job a great deal easier. He can and should continue to seek other truths, no matter which side of the debate is offended by them, but now he no longer has to reject the consequences of those clear observations that disagree with the "accepted" dogmas of someone else. He can see whatever his eyes behold, and conclude whatever the facts lead to, without passing everything through the filter of conformity to the orthodoxy of the day. He will finally be free to find the real truth, whatever it may be.

Such a happy outcome is good and just, because the scientists themselves are the ones who, possibly without meaning to, are engaged in opening the world to these miracles after three hundred years of the Dark Ages of "Enlightenment".

About the Author

Burton Brodt is a retired chemical engineer and R&D manager who holds several patents in his field of process research. He now works as a high school track coach and substitute teacher. At one time he thought about becoming a professional writer, but decided he preferred to eat. He is a recovered atheist.

He has four super children and ten grandchildren, all scattered across the warm zone from Jacksonville to Honolulu, and two equally great step children, in Illinois and Delaware.

Brodt has called several places home, including Chicago; Jacksonville, Florida; LaPlace and New Orleans, Louisiana; Louisville, Kentucky; Clear Lake, Texas; and Wilmington, Delaware. He now lives in Easton, Maryland with his second wife Gail, two friendly horses, and two eccentric cats. He is still recovering from the loss of his friend Zep, the German Shepherd.

Other books by Brodt are:

Four Little Old Men, A (mostly) True Cajun Tale.

The Man Who Flew the Pepsi Sign and Other Stories

Misadventures of a Sporting Life: Blunders, Odd Occurrences, and Hair-Breadth Escapes

The Monkey Wars: The Battle Between Darwinists and Creationists

Ain't Got Anything But: Surviving Egypt

Confessions From the Shark: A Voyage from Hell

Engineers and Scientists: Achieving Success in Industry

Kevlar® Technology (proprietary)

Advances in Neoprene Technology (proprietary)

www.ingramcontent.com/pod-product-compliance
Lightning Source LLC
Chambersburg PA
CBHW051540170526
45165CB00002B/806